Contents

Introduction

The Folens **Specials! Science** series meets the requirements for the Key Stage 3 National Curriculum in the United Kingdom, and is compatible with the Schemes of Work published in England by the Qualification and Curriculum Authority (QCA) and the Scottish 5–14 Guidelines.

The new Key Stage 3 **Specials!** series, has been developed in response to extensive research, which identifies how to aid students with lower reading abilities. Its photocopiable format with easy to follow activity sheets, accompanied by a detailed set of teacher's instructions will enable students to make the most of their learning experience. All books in the Folens **Specials!** series have been assessed through research carried out within the schooling community.

Chemical reactions, materials and particles contains the chemistry topics required at Key Stage 3, and has been completed for students with a reading age of between six and nine. The book is divided into ten sections, containing photocopiable activity sheets, which can be used by individual students or as part of a group activity. Practical activities are included in most units, where students are encouraged to work in small groups using writing frames to plan their own experiments. Teachers can make the most of photocopiable Activity sheets, which can be used by individual students or as an excellent aid for group projects.

Teacher's notes are provided for the teacher as guidance when using **Activity sheets**. They comprise a set of learning **Objectives,** which detail the skills students will acquire in each individual unit. **Prior knowledge** has been set out, referring to scientific knowledge, which students will already be familiar with, in order that they can complete each unit successfully. **Background** information has also been provided for each unit, and gives an overview of the individual topics, highlighting how different pages can be linked together. In addition to the **Activity sheets,** a **Starter activity** has been included to introduce each unit or relate it to a previous topic.

Activity sheets are intended to be taught consecutively and contain a variety of tasks, including sequencing work, filling in the gaps, constructing posters, as well as using ICT to prepare individual presentations and practical sessions. A suggested **Plenary**, which can be found within the accompanying **Teacher's notes**, can be used to recall key points and important terminology within each section.

An **Assessment sheet** which can be found at the end of the book, is intended to highlight the student's progress, whereby students can identify successful tasks they have completed during the course of the book. At this stage, students will be able to provide examples of their own work to support their answers. Assessment sheets are a useful tool for teachers to determine the level at which each student is working.

Look out for other titles in the Folens **Specials! Science** series, which include:

- Life processes and the environment
- Energy, electricity and movement
- Scientific investigation, plants, rocks and outer space.

© Folens

Acids and alkalis

Objectives

- Know that acids and alkalis are an important group of chemicals, which we encounter everyday
- Be able to follow Instructions and work safely in a laboratory environment
- Know what an indicator is
- Be able to use Universal indicator to determine the pH values of different substances
- Understand neutralisation and be able to provide examples

Prior knowledge

Students should know how to behave correctly in a laboratory and be aware of the possible dangers inherent in a practical lesson. Students should also be familiar with the different pieces of laboratory equipment and know how to use them correctly and safely.

QCA link

Unit 7E acids and alkalis

NC links

Sc3 Materials and their properties 3d, 3f

Scottish attainment targets

Environmental studies – Science – Earth and space
Strand – Changing materials
Level E

Background

Acids and alkalis are chemicals commonly used at home and in the laboratory. Many people think that acids are extremely dangerous and can burn you. Some acids can, but most are not hazardous and make an important contribution to our diet, our health and our cleanliness. Alkalis react with acids and cancel them out. Contrary to popular opinion, many strong alkalis are caustic and can burn living tissue.

Starter activity

Ask different students to come up to the board. Instruct them to write down any word, which they associate with 'acids'.

Activity sheets

'Acids and alkalis' introduces students to different everyday substances that belong in this group of chemicals. Provide students with a word bank to complete the introductory text. Tell students to then work in pairs to complete the activity which follows. Students should put the substances, which appear on the sheet, into two groups, acids and alkalis. Once they have completed this, ask students to colour all the acids in red and all the alkalis in blue. Students should then match the start of each sentence provided with its ending. Remind students that a sentence always begins with a capital letter.

'Hazard signs' highlights some examples of ordinary products, which could be dangerous if misused. Ask students to look at the hazard signs provided and guess what they mean. Instruct students to then cut out and match the signs with the meanings.

'Universal indicator' informs students how to link indicator colour change with the pH of some common substances. Provide students with an enlarged A3 copy of the pH scale provided, advising them to place each product in its correct position on the scale, according to its pH value. Tap water is provided as an example.

'Acid or alkali?' Following on from the previous activity sheet, ask students to study the table provided and name each substance according to the experimental results. Once students have completed this task, encourage them to visualise the results, by colouring in the test tubes provided. Ask students to think about the products they might find at home. Encourage students to identify which of these might be classed as acids and which can be classed as alkalis.

'Neutralisation' introduces the students to everyday situations, where a knowledge of chemistry might be put to practical use. Students should use the word bank provided to complete the text.

'Indigestion remedies'. Choose three different indigestion remedies for the students to test. Label them Antacid 1–3. Students should carry out an experiment to determine how well different indigestion remedies might work according to their pH levels. Encourage students to record their observations.

Plenary

Encourage students to test their knowledge by showing the class different hazard signs and asking them to write down their meanings.

Acids and alkalis

☞ Complete the text below, using a word bank provided by your teacher.

> Acids and alkalis are important c _ _ _ _ _ _ _ _ _ . We come into contact with them e _ _ _ _ day both at h _ _ _ and in school.

☞ Write down five things you know about acids. You can work in pairs.

_______________ _______________ _______________ _______________ _______________

☞ Below are the names of lots of different acids and alkalis. You will already know some of them and others will be new to you. Arrange the boxes below into two columns. Label one column **Acids** and the other **Alkalis**. Colour all the acids in red and colour all the alkalis in blue.

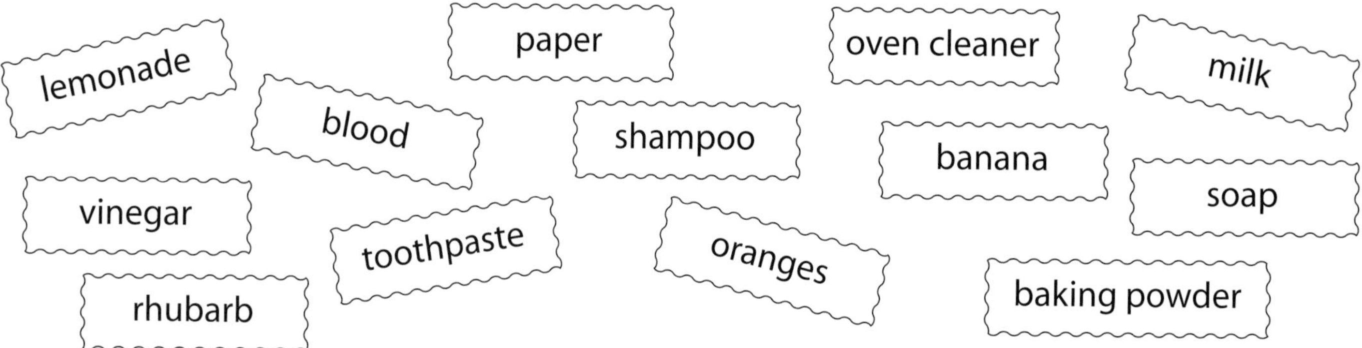

☞ Study the boxes below. Match the start of each sentence with its ending.

Some acids like the sulphuric acid found in car batteries are	an acid.
We use an alkali to	taste.
The vinegar we put on our chips is	very, very dangerous.
Alkalis taste a bit like	clean our teeth.
Acids have a sour	soap.

 © Folens (copiable page)

Hazard signs

☞ Some of the a _ _ _ _ and alkalis we use at school are d _ _ _ _ _ _ _ _ _.

Never taste any chemicals in the laboratory!

☞ Even some of the chemicals we use at home can be dangerous if not handled carefully. Your teacher will show you some household products, which you might find in your home. Look at the labelling on the containers. Can you see any of the signs printed below?

Now, cut out the signs and the word boxes provided below. Try to match each hazard sign with its meaning. Next to each hazard sign write down the name of a product displaying that sign.

Corrosive
Attacks and destroys living tissues.
Will burn your skin.

Irritant
Can cause skin to go red and blister.

Toxic
Poisonous!

☞ Study the information below. Always remember this when handling acids or alkalis.

What should you do if you accidently spill an acid or alkali:

- On your skin?

Hold under running water immediately. Tell your teacher.

- On your clothes or books?

Tell your teacher.

Universal indicator

☞ Complete the text below, using a word bank provided by your teacher.

> Universal indicator is a m _ _ _ _ _ _ _ of different c _ _ _ _ _ _ _ _ chemicals, which change colours in acid and alkali solutions. The different colours of universal indicator tell us if an acid or a _ _ _ _ _ _ is s _ _ _ _ _ _ or weak.

☞ Your teacher will provide you with a coloured universal indicator chart to complete the next activity.

On an A3 piece of paper, copy out the pH chart below and colour it in. Now write each product listed below in its correct position on the chart. Once you have finished, you may wish to add in illustrations.

pH
0
1
2
3
4
5
6
7
8
9
10
11
12
13
14

acid ↑ ↓ alkali

Baking powder pH8.5

Lemon pH4

Vinegar pH2.5

Tap water pH7

Soap pH9

Fertilizer pH8

Fizzy lemonade pH5

Indigestion tablets pH10

Hydrochloric acid pH2

Sugar pH7

Shampoo pH8

Car battery acid pH0

Paper pH9

Milk pH7.5

Toothpaste pH8

Drain cleaner pH14

Orange squash pH6

© Folens (copiable page)

Acid or alkali?

Some of the products we use at home are acids and some are alkalis. Paula and Niamh wanted to find out which of the different products they used at home were alkalis or acids. The substances tested were tap water, vinegar, toothpaste, dilute drain cleaner and lemonade.

☞ Study Paula and Niamh's results below. Can you name each solution according to its colour, once universal indicator has been added?

Solution	Colour before adding universal indicator	Colour after adding universal indicator	Name of solution
A	Clear green	Purple	
B	Clear, colourless	Green	
C	Clear, brown	Red	
D	Clear, colourless	Orange	
E	Cloudy, white	Blue	

Now, colour in the test tubes below to show Paula and Niamh's results. Add in the name of each solution.

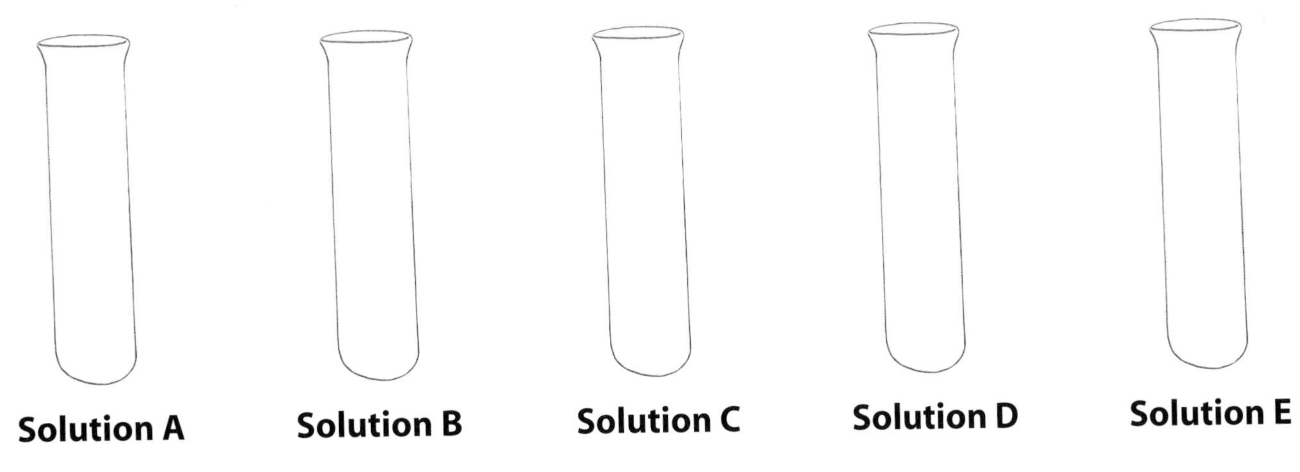

☞ Write down two more acids and two more alkalis, which we use at home?

Acids ______________ ______________

Alkalis ______________ ______________

© Folens (copiable page)

Neutralisation

When an acid and an alkali are mixed together, a chemical reaction happens. The acid and the alkali cancel each other out. We call this **neutralisation**. Neutralisation can be very useful.

☞ Complete the text below, using the word bank provided (at the bottom of the page).

A bee's sting contains an acid, that's why it hurts!

How can you stop the p _ _ _ from a bee sting?

Rub it with an a _ _ _ _ _ like toothpaste or bicarbonate of soda.

Wasp stings are alkali.

What would you use to neutralise a w _ _ _ sting?

I would use v _ _ _ _ _ _ to neutralise a wasp sting because it is an _ _ _ _.

What about indigestion?

Indigestion is caused by too much acid (h _ _ _ _ _ _ _ _ _ _ _ a _ _ _) in the stomach. Your s _ _ _ _ _ _ needs acid to help break down (d _ _ _ _ _) proteins but t _ _ much stomach acid can cause the pain of indigestion. You can buy medicines called antacids which n _ _ _ _ _ _ _ _ _ the effect of the stomach acid.

Do you know the names of any antacids? What type of chemical do they contain?

If a farmer's soil is too acidic, the crops do not grow very well. If he adds lime to the soil this will make it l _ _ _ acid. Lime is an a _ _ _ _ _.

Word bank

too pain digest less alkali stomach alkali
hydrochloric acid neutralise wasp vinegar acid lime

 © Folens (copiable page)

Indigestion remedies

Indigestion is caused when too much acid is made by the stomach. Indigestion remedies are antacids which neutralise the acid and stop the indigestion.

☞ Which is the best indigestion remedy? Use the instructions below to carry out your experiment. Record your results in the table below.

Plan *(What to do)*:

- Measure 50ml of stomach acid (hydrochloric acid) into a beaker using a measuring cylinder.
- Add 5-6 drops of universal indicator. Record the pH.
- Crush some of the indigestion remedy (antacid 1).
- Add three spatulas full of the indigestion remedy to the beaker and stir.
- Is there any fizzing? (record on your results chart).
- When the reaction has stopped, record the colour of the universal indicator and the (pH).
- Did the acid fizz? Write down yes or no next to each one.

☞ Repeat these steps using the other two indigestion remedies. Which antacid remedy was the best?

Observations and results

Antacid	Colour and pH at the beginning	Colour and pH at the end
1		
2		
3		

© Folens (copiable page)

Simple chemical reactions

Objectives

- Understand the terms 'chemical' and 'physical' reaction, and be able to give examples of each
- Develop observational skills
- Know the test for hydrogen gas
- Know the test for carbon dioxide gas
- Recognise that burning requires oxygen

Prior knowledge

Students should already have experience which covers working in the laboratory. Ensure students know which equipment to use and are able to follow simple written instructions.

QCA link

Unit 7F Simple chemical reactions

NC links

Sc3 Materials and their properties 1f, 2h, 2i, 3a, 3e

Scottish attainment targets

Environmental studies – Science – Earth and space
Strand – Changing materials
Level E, F

Background

When a chemical change takes place, new substances are always produced. Chemical changes are difficult to reverse. Sometimes gases are produced during a chemical reaction and these can be identified using the appropriate tests. Burning is a common chemical reaction, resulting in the production of new chemicals as well as releasing stored energy from the fuel as heat (thermal) energy.

Starter activity

Encourage students to look at two different types of reaction. Ask a student to strike a match and with it, light a candle. **Students must be supervised carefully when lighting matches**. Students should then discuss the following ideas: *What has happened? Can I use that match again?* Repeat the exercise by asking another student to blow up a balloon and then let it down. Students should ask themselves, *Can I use that balloon again?*

Activity sheets

'What is a chemical reaction?' introduces students to the words, **chemical** and **physical** reactions. Ask students to complete the introductory text using a word bank. Students should then use the information provided, which looks at the process of making a cake, to fill in the missing words. Ask students to make similar picture equations for making a cup of tea and some toast.

'More chemical reactions' follows on from the previous activity sheet. Tell students to follow the experiments provided, which will help them to observe different chemical reactions. Students should record their observations within the chart provided.

'What gas is this?' reintroduces hazard signs. Students should watch a demonstration of tasks 1 and 2 before completing the activity. Ask students to work in pairs to make their own gas and test it; they should think about what gas they have made. Students should now have the sufficient information to label the diagram, which follows.

'Acids and carbonates' offers a practical lesson for testing different gases. Ask students to observe the experiment provided, instructing them to complete the questions, which follow it. Students should then complete the second half of the activity sheet by crossing out the incorrect words or phrases.

'Fuels'. This activity sheet explains the conditions needed for a fuel to burn. Ask students to complete the sentences surrounding the fire triangle. Students should use these answers to help them to complete the second activity, 'putting out a fire'. Encourage students to use the activity sheet to make a safety poster for bonfire night.

Plenary

Provide students with pre-made cards including different phrases or words, such as; *melting ice, boiling water, lighting a candle, blowing up a balloon* and *making a cup of coffee*. Ask students to place the cards into two piles. One pile should signify chemical reactions, the other should represent physical reactions.

© Folens

What is a chemical reaction?

☞ Complete the introduction below.

Changes are happening around us all the time. Some are c _ _ _ _ _ _ _ _ changes and others are p _ _ _ _ _ _ _ changes.

☞ Identify which of the reactions below are chemical changes and which are p_ _ _ _ _ _ _ changes.

What type of change is it?

Blowing up a balloon
Is a ________________ change.

Making a cup of tea
Is a ________________ change.

Melting ice cream
Is a ________________ change.

A burning match
Is a ________________ change.

☞ Complete the text below. Then, use the text to help you to create word equations for making a cup of tea, and making toast. Use drawings to illustrate your work.

A c _ _ _ _ _ _ _ _ reaction has happened when a n _ _ substance is made. We call this new substance the p _ _ _ _ _ _. The substances which it is made from are called the r _ _ _ _ _ _ _ _ _.

☞ Here is an example of a chemical reaction:

flour
+
eggs
+
sugar
+
milk

Mix together and cook in the oven

cake

When the flour, eggs, sugar and milk (reactants) are mixed together and then baked (h _ _ _ _ _) in the oven to make a cake (p _ _ _ _ _ _), it is n _ _ possible to get the original ingredients (reactants) back again.

More about chemical reactions

☞ Complete the introduction below.

I already know that something n _ _ is m _ _ _ during a c _ _ _ _ _ _ _ reaction.

☞ Sometimes the product and the reactants look the same. How can we tell if there has been a chemical reaction? Below are some experiments that will help you to find out. Complete the experiments, checking with your teacher before you begin. Record your results in the spaces provided.

Experiment	What did I see?	What did I hear?	What did I feel?
• Put 20cc of lemon juice (colourless liquid) into a test tube. • Add a spatula of bicarbonate of soda (a white powder).			
• Put 20cc of hydrochloric acid (colourless liquid) into a test tube. • Add a piece of magnesium ribbon (grey metal).			
• Put 20cc of copper sulphate solution (clear blue liquid) into a test tube. • Add an iron nail (grey metal).			

☞ I know that there has been a chemical change if:

- I see lots of b _ _ _ _ _ _
- I hear f _ _ _ _ _ _
- I see a different c _ _ _ _ _ _ _ product
- I feel the test tube get w _ _ _
- I feel the test tube get c _ _ _ _ _ .

© Folens (copiable page)

What gas is this?

This is the hazard sign for a c _ _ _ _ _ _ _ _ _ substance. Acids are corrosive; they eat away (corrode) other substances such as your clothes, your skin and even metals.

Corrosion is a chemical reaction

Metal + acid ⟶ metal salt + colourless gas

R _ _ _ _ _ _ _ _ _ P _ _ _ _ _ _ _ _

☞ In pairs, complete the tasks, which follow. Check with your teacher, before beginning each task.

Task 1

Making and collecting the colourless gas

- Put a piece of magnesium ribbon in a test tube.
- Stand the test tube in a rack.
- Carefully add the acid so that it just covers the metal.
- Very quickly put another test tube upside down over the mouth of the one in the rack.
- The colourless gas will rise upwards and collect inside the second test tube.

Task 2

Testing the gas

- Have a lighted splint ready.
- Lift the second test tube straight upwards.
- Put the lighted splint inside the mouth of this test tube.
- What did you hear?

☞ Watch the pop test demonstration by your teacher. Label the diagram opposite.

This is called 'The squeaky pop test'. Only hydrogen gas burns with a squeaky pop.

© Folens (copiable page)

Acids and carbonates

The chemical name for chalk is calcium carbonate. Carbonates always contain the elements carbon and oxygen. When an acid is added to a carbonate, a colourless gas is made.

☞ Your teacher will show you what happens when an acid is added to a carbonate. Study the experiment carefully to see what happens when a carbon dioxide gas is made. Then complete the questions.

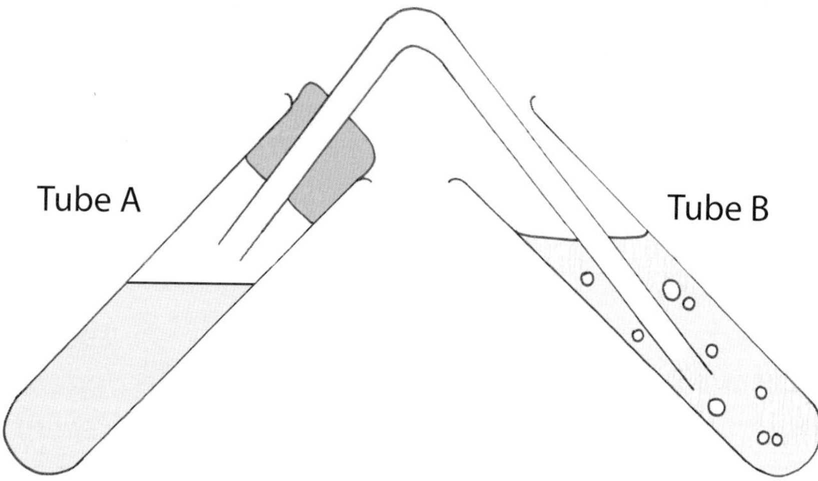

1. What is in tube A? ____________________
2. What is in tube B? ____________________
3. What gas is in the bubbles? ____________________

☞ Now, cross out the incorrect words or phrases.

- When an acid is added to a carbonate it always ***fizzes/changes colour***.
- The gas in the bubbles is called ***oxygen/carbon dioxide/sulphur dioxide***.
- Carbon dioxide is a ***coloured/colourless*** **gas,** that has a ***strong smell/no smell***.
- Carbon dioxide always contains ***carbon/hydrogen/oxygen/nitrogen*** atoms.
- We test for carbon dioxide gas using ***tap water/limewater***.
- When the gas bubbles pass through limewater it becomes ***blue/milky white***.
- This shows us that the bubbles are made of ***oxygen/carbon dioxide*** **gas**.

 © Folens (copiable page)

Fuels

A fuel is something that burns and gives off heat. Fuels can be solids, liquids or gases.

☞ Name five fuels. ______________ ______________ ______________

______________ ______________

☞ What do fuels need to burn? Complete the text below.

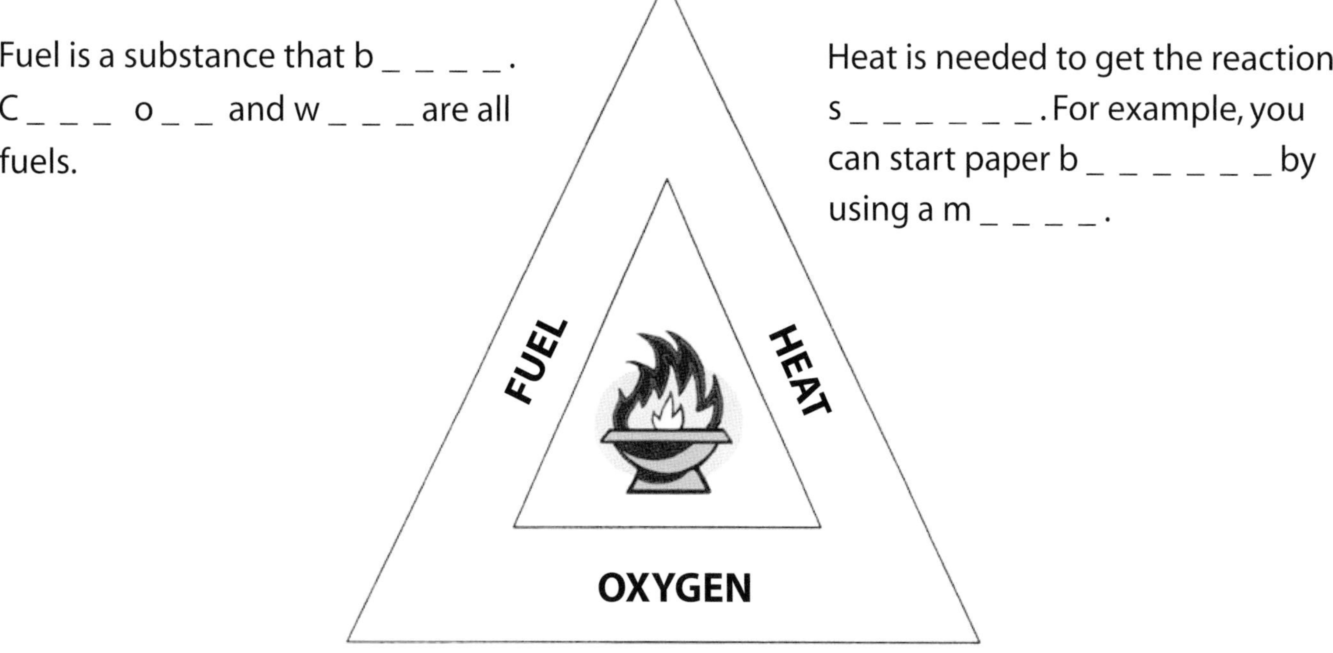

Burning needs oxygen. Without o _ _ _ _ _ a fire will go o _ _.

☞ Use what you have learnt to complete the text below.

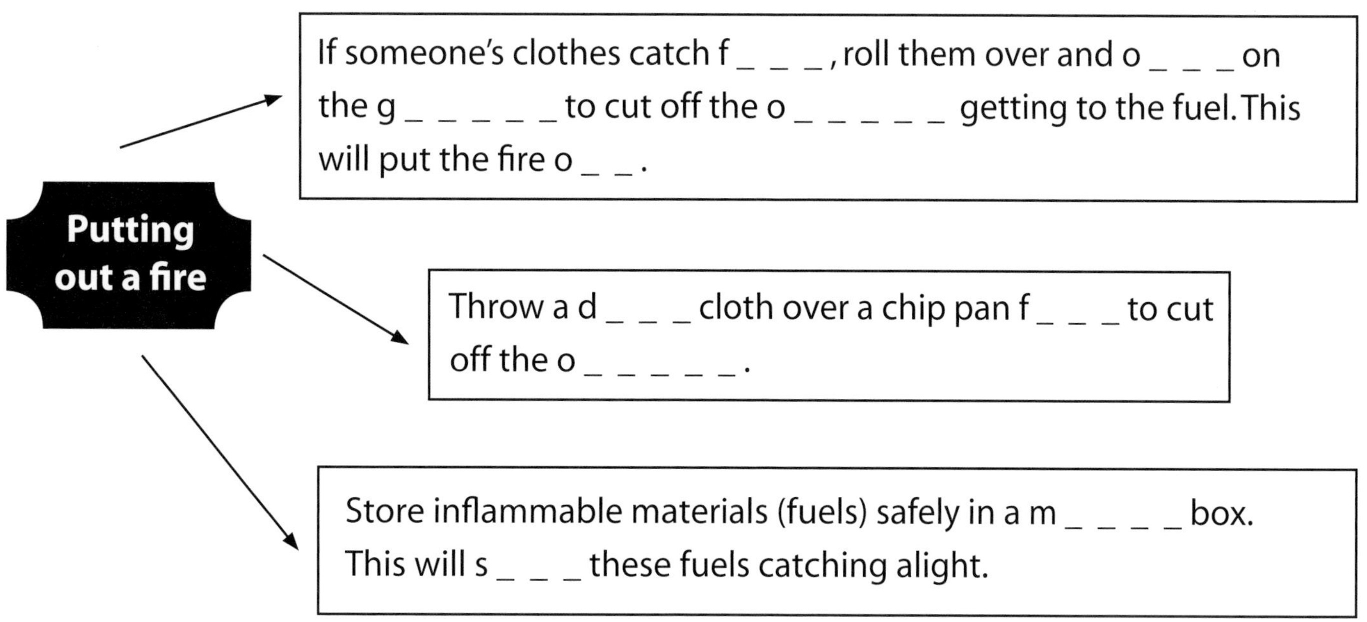

Particle model of solids, liquids and gases

Objectives

- Know that everything is made up of particles
- Understand the way particles are arranged in solids, liquids and gases
- Identify how the arrangement of particles affects the properties of solids, liquids and gases
- Understand how particles move through the air
- Be able to interpret experimental results

Prior knowledge

Students may already know that solids, liquids and gases are made up of small particles. Students should be able to name different examples of solids, liquids and gases.

QCA link

Unit 7G Particle model of solids, liquids and gases.

NC link

Sc2 Materials and their properties, 1b

Scottish attainment targets

Environmental studies – Science – Earth and space
Strand – Materials from Earth
Level E

Background

Everything is made up of small particles, too small to be seen without an extremely powerful microscope. Particles are arranged in a definite way and the arrangement of these particles determines the way in which solids, liquids and gases behave.

Starter activity

Nominate groups of students to represent particles in either solids, liquids or gases. Solid particles should stand close together, swaying gently from side to side. Instruct the students representing liquid particles to stand quite close together and move around within a prescribed space. Those representing gases can move all around the room in any direction. Ask each group to act out its role to the rest of the class.

Activity sheets

'Particles'. Ask students to complete the introductory text, using the word bank provided. Instruct students to copy out the table, which follows, so it fills an A4 sheet. Students should then use the phrases provided to complete the table. Encourage students to extend this activity by drawing an example of a liquid, a gas and a solid.

'How are particles arranged?' Students should complete the text using the word bank provided. Now provide students with sticky-backed dots, asking them to place the dots in each box to show how the particles are arranged in a solid, a liquid and a gas.

'Gases on the move' looks at an experiment used to show diffusion of gas particles. Results have already been provided on sheet 2. Ask students to complete the introduction. You may wish to provide students with a word bank. Students should then use the information on the activity sheet to answer the questions, which follow. Instruct students to then study the results on sheet 2, and to plot a simple bar graph on squared paper. Students should answer the questions, which follow, using the information from this activity as well as information from a previous topic (Acids and alkalis).

Plenary

Divide the board up into three columns; **solids**, **liquids** and **gases**. Ask students to use different coloured board markers to write one example in any of the columns.

 © Folens

Particles

Word bank

solid small shape liquid
destroyed size gas seen

☞ Complete the sentences below, using the word bank.

- A particle is a very, very s _ _ _ _ piece of something.
- Particles are too small to be s _ _ _ with your eye or even with a school microscope.
- Particles can not be d _ _ _ _ _ _ _ _ _.
- Particles in every substance have their own special fixed s _ _ _ _ and s _ _ _.
- The arrangement of particles in a substance decides whether it is a s _ _ _ _, l _ _ _ _ _ or g _ _.

☞ Copy the table below onto an A4 sheet, leaving room for illustrations. Study the circles below the table. Put each circle into its correct place in the table.

Solid	Liquid	Gas
brick	MILK milk	air

Changes shape to fit the container

Particles move around in all directions

Particles are packed close together

Particles are not tightly packed and can move a little

Has a fixed shape

No force between particles

Large pull force between particles

Particles are far apart

How are particles arranged?

☞ Complete the text below, using the word bank provided. Now, use coloured sticky dots to show how particles are arranged in a solid, a liquid and a gas. Your teacher will help you.

Word bank

around directions close move
apart solid together vibrate
gas move far

In a s _ _ _ _ the particles are packed very c _ _ _ _ together. The particles in a solid do not m _ _ _ but can v _ _ _ _ _ _ in the same spot.

In a liquid, the particles can m _ _ _ around in all d _ _ _ _ _ _ _ _ _ but they are still packed closely t _ _ _ _ _ _ _.

The particles in a g _ _ are f _ _ apart and move a _ _ _ _ _ quickly.

 © Folens (copiable page)

Gases on the move sheet 1

☞ Complete the text below.

I know that in a gas the p _ _ _ _ _ _ _ _ _ are not held c _ _ _ _ _ _ _ together. They can m _ _ _ around and s _ _ _ _ _ out to fill their container.

John and Nabeel designed an experiment to see if gases really do move. They used a gas called ammonia. Ammonia is an invisible gas. Ammonia makes Universal Indicator paper change to a blue/purple colour.

☞ The diagram below shows how they carried out their experiment. Their results are provided on sheet 2. Before reading their results, study the diagram and answer the questions, which follow.

Small piece of damp Universal Indicator paper

Upside down boiling tube

White tile

Cotton wool ball soaked in ammonia

Grease around opening of boiling tube

☞ Why did the students put grease around the neck of the boiling tube?

____

☞ Why did the students use damp Universal Indicator paper?

____

Gases on the move sheet 2

☞ John and Nabeel's results are provided below. Use their results to plot a simple bar graph onto squared paper.

Distance from bottom of boiling tube (cms)	Time taken to change colour (seconds)
2	3
4	20
6	45
8	92
10	190

☞ Now, answer the questions below. You may need to refer back to a previous topic to help you.

1. Which piece of indicator paper changed colour first?

____

2. In what order did the pieces of indicator paper change colour?

____

3. What do John and Nabeel's results show about the way gas particles move?

____

4. What type of chemical is ammonia? Look at the colour change to help you.

____

5. If the students had used an acidic gas, what colour change would they have seen?

____

 © Folens (copiable page)

Solutions

Objectives

- Understand that when a solute dissolves in a solvent, it is still there even if we can not see it
- Know what is meant by filtering and be able to carry out the process
- Understand how to separate a solute from a solvent by evaporation
- Identify how to get pure water from a mixture
- Know how to separate a mixture of dyes by using chromatography
- Be able to put each step of an experiment procedure in the correct order

Prior knowledge

Students should be aware that some substances will dissolve in water whilst others will not. Students will need to be able to provide examples of soluble and insoluble substances.

QCA link

Unit 7H Solutions

NC links

Sc3 Materials and their properties, 1b, 1g, 1h, 2a, 2b

Scottish attainment targets

Environmental studies – Science – Earth and space
Strand – Changing materials
Level D

Background

Most of the time it is difficult to tell if water has anything dissolved in it, because it still looks the same; a clear, colourless liquid. Sometimes a colour lets us know that a substance has dissolved in the water. For example, copper sulphate gives a clear, blue solution. We can use different techniques to get back the solute, or to make pure water.

Starter activity

Ask students to make two lists, one of all the substances, which they can think of that will dissolve in water, and a second of all the things which will not dissolve in water.

Activity sheets

'Solutions' asks students to make up key words associated with solutions using the vowels provided. Ask students to match the key words with their meanings. Students will be able to use these words to identify **brine** in the activity, which follows.

'Dissolving sugar in water' introduces the different conditions, which affect the rate at which sugar will dissolve. Ask students to identify the correct words, which describe the dissolving process. Students should then complete the list, which follows. This activity sheet provides a good introduction for a practical lesson investigating factors affecting the rate at which sugar dissolves.

'Separating mixtures' is a sequencing exercise. Ask students to cut out the boxes provided and put them in the correct order. Students should leave enough space between each stage for illustrations. Once the sentences have been placed in their correct order, instruct students to draw a diagram of the apparatus for each stage of the procedure. Teachers may wish to use this activity sheet as a follow on exercise, whereby students carry out a practical experiment using filtering and evaporation to separate the constituents of salt.

'What colour is black ink?' introduces the technique of chromatography. Ask students to cut out each sentence and arrange them in the correct order. Tell students to then use these instructions to design an experiment to separate the colours in a black water-based felt-tip pen.

'How do we separate different mixtures?' Ask students to form pairs. Each pair should complete the boxes provided using a word bank. Ask students to then copy the boxes onto an A3 sheet of paper, enabling them to insert drawings to illustrate each method.

Plenary

Choose one of the key words (for example, chromatography) to start a class game of hangman.

Solutions

☞ Use the vowels provided to complete the keywords associated with solutions.

a e i o u

d _ ss _ lve   s _ l v _ n t   s _ l _ t _ _ n

s _ l _ t _

_ n s _ l _ b l _   s _ l _ b l _   s _ t _ r _ t _ d

☞ Study the table below. Draw lines from each word to its correct meaning.

Word	Meaning
Solution	A liquid that will dissolve something
Solute	A solid mixed with a liquid
Soluble	No more solid can be dissolved
Solvent	Solid that dissolves in a solution
Saturated	Something that can be dissolved in a liquid
Dissolve	Can not dissolve in a liquid
Insoluble	When a solid mixes with a liquid to make a solution.

☞ Use your table to complete the sentences below.

Brine is a mixture of salt and water.

The solvent is ____________________

The solute is ____________________

The solution is ____________________

© Folens (copiable page)

Dissolving sugar in water

☞ Below is a diagram, which explains how sugar dissolves in water. Study the text and cross out the incorrect phrases in italics.

Sugar dissolves ***slower/quicker*** in hot water than in cold water.

Sugar is ***soluble/ insoluble*** in water.

You can make sugar dissolve more ***quickly/ slowly*** in water by using a spoon to stir the water.

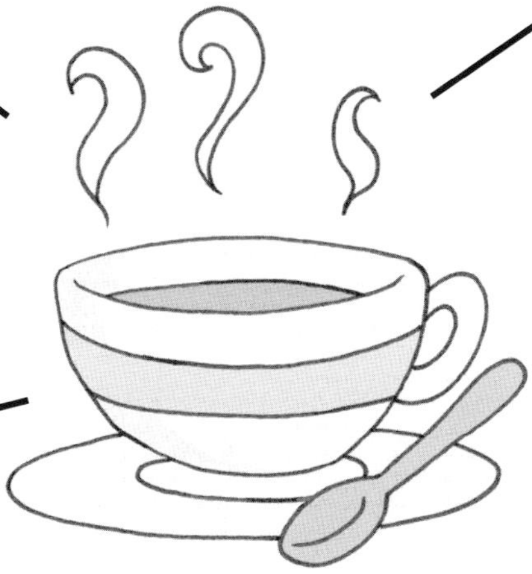

Small ***granules/ large lumps*** of sugar dissolve more ***slowly/ quickly*** in the water.

The ***slower/quicker*** you stir, the sooner all the sugar ***will/will not*** dissolve.

If you keep adding sugar, eventually the solution will become ***satisfied/saturated*** and ***no more/lots more*** sugar will dissolve.

☞ List three of the conditions needed for sugar to dissolve more quickly in water.

- ____________________________________
- ____________________________________
- ____________________________________

Separating mixtures

☞ The boxes below represent an experiment for separating mixtures. Stages of the experiment have been jumbled up. Cut out the sentences, and arrange them in the correct order to create a flow chart. Leave enough room for drawings later.

Pour the rock salt and water mixture through a filter funnel.	Add rock salt to a beaker of water and stir.
The water evaporates and small white crystals of salt are left in the dish.	Clear drops of liquid drip through the filter paper.
The insoluble sand remains on the filter paper.	The clear liquid is poured into an evaporating dish.
The clear liquid is heated until it boils.	Rock salt is a mixture of sand and salt.
Use a pestle and mortar to crush the rock salt.	This clear liquid collects in the beaker.

☞ Now, draw diagrams to represent each stage of the experiment. Here is an example.

Rock salt is a mixture of sand and salt.

© Folens (copiable page)

What colour is black ink?

☞ Complete the text below.

A pure dye contains only o _ _ c _ _ _ _ _ _. Black ink is n _ _ a p _ _ _ dye and it contains m _ _ _ than one colour.

☞ Now, cut out the boxes below. In your work book, place each box in its correct order to create different sentences. Once in their correct order, use the sentences to design an experiment to separate the colours in a black water-based felt-tip pen.

Separating colours

We can separate the different

a different speed across the paper, so all the different dyes in the black

piece of special chromatography paper. Each dye moves at

can see the different colours that make up black ink.

to dissolve the different coloured dyes in the black ink. The dyes are carried across a

ink move at their own speed. The dyes become separated from each other and we

colours in black ink using chromatography. A liquid solvent is used

How do we separate different mixtures?

☞ Copy all the boxes below onto a sheet of A3 paper. In pairs, complete the boxes, using a word bank provided by your teacher. Add drawings to illustrate each method of separation.

How do we separate different mixtures?

S _ _ _ _ _ _ _ _ substances + water
↓
I _ _ + water
↓
Black s _ _ _ _ _ _ _ _
↓
D _ _ _ _ _ _ _ _ _ _ _ _ _
↓
P _ _ _ water

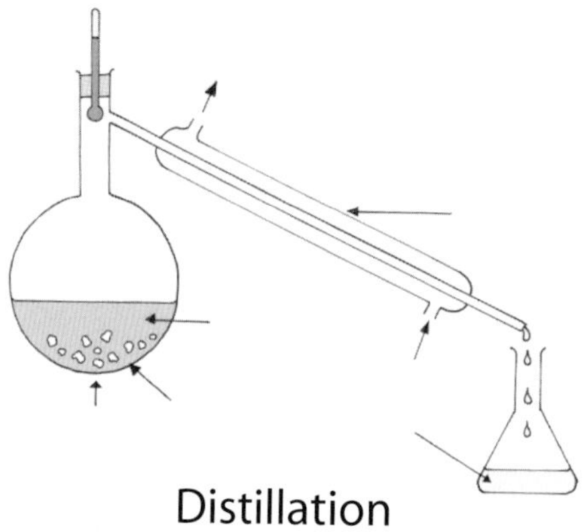

Distillation

Soluble s _ _ _ _ _ _ _ _ _ + water
↓
S _ _ _ + water
↓
Brine s _ _ _ _ _ _ _ _
↓
E _ _ _ _ _ _ _ _ _ _ _
↓
Salt c _ _ _ _ _ _ _ _

Mixture of c _ _ _ _ _ _ _ _ dyes
↓
Black i _ _
↓
C _ _ _ _ _ _ _ _ _ _ _ _ _ _
↓
S _ _ _ _ _ _ _ _ _ the colours

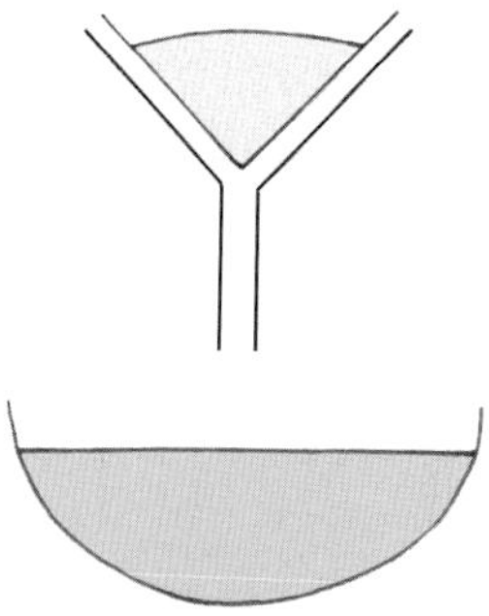

I _ _ _ _ _ _ _ _ _ substances + water
↓
S _ _ _ + water
↓
M _ _ _ _ _ _ _
↓
F _ _ _ _ _ _
↓
S _ _ _

Filtering

© Folens (copiable page)

Atoms and elements

Objectives

- Know that everything is made up from a basic set of bits called elements
- Identify that each element contains only one type of atom
- Understand how the elements are listed in the periodic table
- Know that elements can join together to make compounds

Prior knowledge

Students should know that there are many different types of materials available. These materials are made up of different substances.

QCA link

Unit 8E Atoms and elements

NC links

Sc3 Materials and their properties 1a,1c, 1d,1e, 1f

Scottish attainment targets

Environmental studies – Science – Earth and space
Strand – Materials from Earth
Level E, F

Background

Everything is made from elements. Some materials, like iron, contain only one element. Other materials, like nylon, are made from a compound containing two or more elements. Elements are arranged in the Periodic Table in groups.

Starter activity

Ask students to write a list of as many different materials they can think of. Then, circle all those that contain only one substance.

Activity sheets

'Elements' introduces students to the idea that some elements are solids, whilst some are liquids and others are gases. Ask students to complete the introductory text. Then using the example provided, ask students to add a diagram to each box in order to illustrate that element.

'Periodic table' is a resource sheet which students should use in future lessons. Show students where the metals are on the table, and ask them to colour the metals lightly in blue. Inform students that the rest of the elements are non-metals. Ask students to colour all non-metals in yellow. Remind students to colour lightly, so as not to obscure the letters.

'More about the periodic table' explains what the periodic table is. Ask students to use their copy of the periodic table to answer the questions. Students should continue to use their periodic table as a reference to complete the table provided.

'Compounds' uses the word CHEMISTRY to illustrate that each of the letters can be used in different combinations to make a new word. Ask students to make as many words as they can from CHEMISTRY. Students should then identify all words which make up elements by colouring them in blue, and all words which make up compounds, by colouring them in green.

Plenary

Provide students with pre-made cards including names of different elements. Ask the students to make two piles; one made up of metals and another made up of non-metals. Tell students to use their periodic table to check their answers.

Elements

☞ Complete the text below.

All materials are made up of substances called elements. There are over 100 elements. Some materials are made of only o _ _ element. Most of the materials we use in our daily lives are made up of t _ _ or more d _ _ _ _ _ _ _ _ elements joined together.

☞ Study the boxes below. Draw a diagram inside each box to show what each of these elements is used for. One diagram has already been drawn for you.

Copper is an element. In your house, the wire inside all the e _ _ _ _ _ _ _ _ _ _ circuits is made of the element copper. The pipes that carry w _ _ _ _ around your house are made of copper.

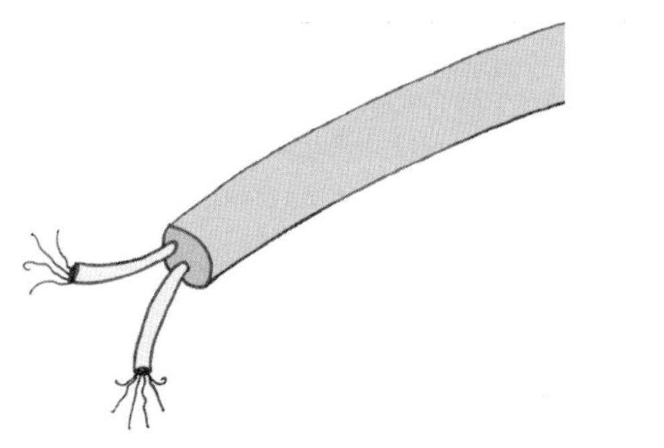

Iron is another element. What do you know that is only made of iron?

Not all elements are solids. Mercury is a liquid element. Where is mercury used?

Some elements are gases. Neon is a gas. Where is neon used?

 © Folens (copiable page)

Periodic table

☞ Colour the metals in blue and the non-metals in yellow.

1.	2.											3.	4.	5.	6.	7.	0.
							H 1										He 2
Li 3	Be 4											B 5	C 6	N 7	O 8	F 9	Ne 10
Na 11	Mg 12											Al 13	Si 14	P 15	S 16	Cl 17	Ar 18
K 19	Ca 20	Sc 21	Ti 22	V 23	Cr 24	Mn 25	Fe 26	Co 27	Ni 28	Cu 29	Zn 30	Ga 31	Ge 32	As 33	Se 34	Br 35	Kr 36
Rb 37	Sr 38	Y 39	Zr 40	Nb 41	Mo 42	Tc 43	Ru 44	Rh 45	Pd 46	Ag 47	Cd 48	In 49	Sn 50	Sb 51	Te 52	I 53	Xe 54
Cs 55	Ba 56	La 57	Hf 72	Ta 73	W 74	Re 75	Os 76	Ir 77	Pt 78	Au 79	Hg 80	Tl 81	Pb 82	Bi 83	Po 84	At 85	Rn 86
Fr 87	Ra 88	Ac 89	Rf 104	Db 105	Sg 106	Bh 107	Hs 108	Mt 109	Ds 110								

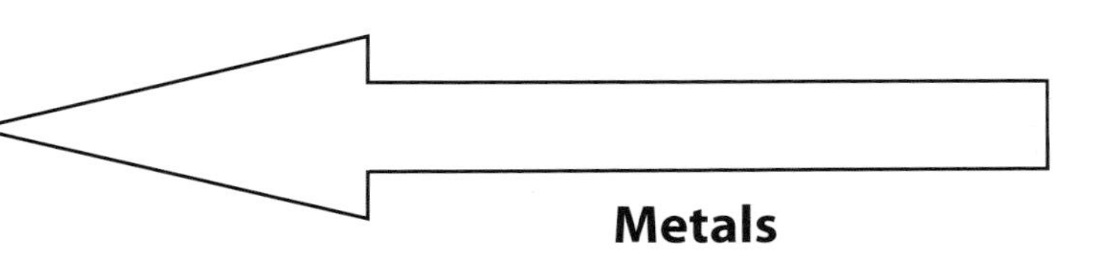

Non-metals

© Folens (copiable page)

More about the periodic table

The periodic table is the way that scientists classify all the elements. The elements are arranged in order of increasing size. The periodic table starts with element number 1, Hydrogen, which is the lightest element. The vertical columns are called groups. Across the top of the periodic table are the numbers of all the eight groups. The elements in each group behave in a similar way and have similar properties. The horizontal rows are called periods. As you move from group 1 across the table to group 0 the elements gradually become less active.

☞ Use your copy of the periodic table to help you to answer the questions below.

1. Where are the metals?

____

2. Are there more metals or more non-metals shown on the periodic table?

____

3. Name two metals which can be found in group 1.

____

4. Are the elements in group 0 more, or less active than those in group 3?

____

☞ Now, complete the table below.

Element	Group	Metal or Non-metal
Sodium		
Carbon		
Neon		
Aluminium		
Chlorine		
Oxygen		
Lithium		
Magnesium		
Sulphur		
Potassium		

 © Folens (copiable page)

Compounds

CHEMISTRY

☞ Use the letters from the word CHEMISTRY to make as many different words as you can. Write down your answers in your work books.

If each of the letters in the word CHEMISTRY were to represent one element you can see just how many new combinations of the letters you can make.

Similarly, elements can also join together to make new combinations called compounds. A compound is made up of two or more different elements joined together to make a new substance.

☞ Study the different substances below. Colour all the elements in blue and all the compounds in green.

Lead

Water

Magnesium oxide

Paper

Salt

Rubber

Aluminium

Helium

Gold

Sugar

Hydrochloric acid

Potassium

Petrol

Wood

© Folens (copiable page)

Compounds and mixtures

Objectives

- Know that a compound is made when two or more elements have joined together in a chemical reaction
- Understand that a compound has different properties to those elements that it is made from
- Identify the chemical symbols for some common elements
- Know what a mixture is and be able to identify some examples
- Use experimental methods to find out if a substance is pure, or if it is a mixture

Prior knowledge

Students should know that everything is made up of elements and that elements can join together chemically to make a compound. Students will be able to find the different elements on the periodic table and know where metals and non-metals are located on the table. To achieve these objectives, students will also need to refer to their Periodic Tables from 'Atoms and elements'.

QCA link

Unit 8F Compounds and mixtures

NC links

Sc3 Materials and their properties, 1a,1c,1d,1e,1f,1g

Scottish attainment targets

Environmental studies – Science – Earth and Space Strand – Changing materials
Level E, F

Background

Different compounds are made up of different numbers of elements. Each element has its own symbol and each compound has its own formula. For example, oxygen gas only contains oxygen atoms, while water contains one hydrogen atom and two oxygen atoms.

Starter activity

Provide students with pre-made elements cards from the previous plenary ('Atoms and elements'). Ask students to use their Periodic Table to find symbols for the different elements.

Activity sheets

'Making compounds' uses an illustrated word equation, which shows how two elements combine to form a compound. Provide students with a word bank to complete the introductory text. Ask students to make their own illustrated word equations for the reactions provided, using the example as a guide.

'Element or compound?' Ask students to study the chemicals provided. Students should then draw arrows between each chemical and its correct column. Instruct students to complete the table, which follows. Students may wish to refer to their Periodic Table as a guide.

'The atoms in a compound' builds upon the previous activity ('Element or compound?') Ask students to draw diagrams to represent the atoms in each of the molecules listed, using the water molecule as an example. Encourage students to identify which of the molecules are compounds.

'What is a mixture?' Ask students to create their own word bank to complete the activities, which follow. Students should check their answers by conferring with fellow students.

'Pure substances'. Students should complete the introductory text in pairs. Students may need to be provided with a word bank to help them. Ask students to design an experiment to see whether adding salt to water alters its boiling point. Check the students' methods before instructing them to carry out their experiments. Students should then complete the concluding text using what they have learnt.

Plenary

Using the pre-made elements cards from the starter activity, ask students to write down symbols without referring to their periodic table. Identify how many symbols each student could remember.

© Folens

Making compounds

☞ Complete the text below. Your teacher will provide you with a word bank to help you.

A c_ _ _ _ _ _ _ _ is made when a _ _ _ _ of t_ _ or more e_ _ _ _ _ _ _ _ have joined t_ _ _ _ _ _ _ _ in a c_ _ _ _ _ _ _ _ reaction.
The compound can look d_ _ _ _ _ _ _ _ _ to the elements from which it was made.

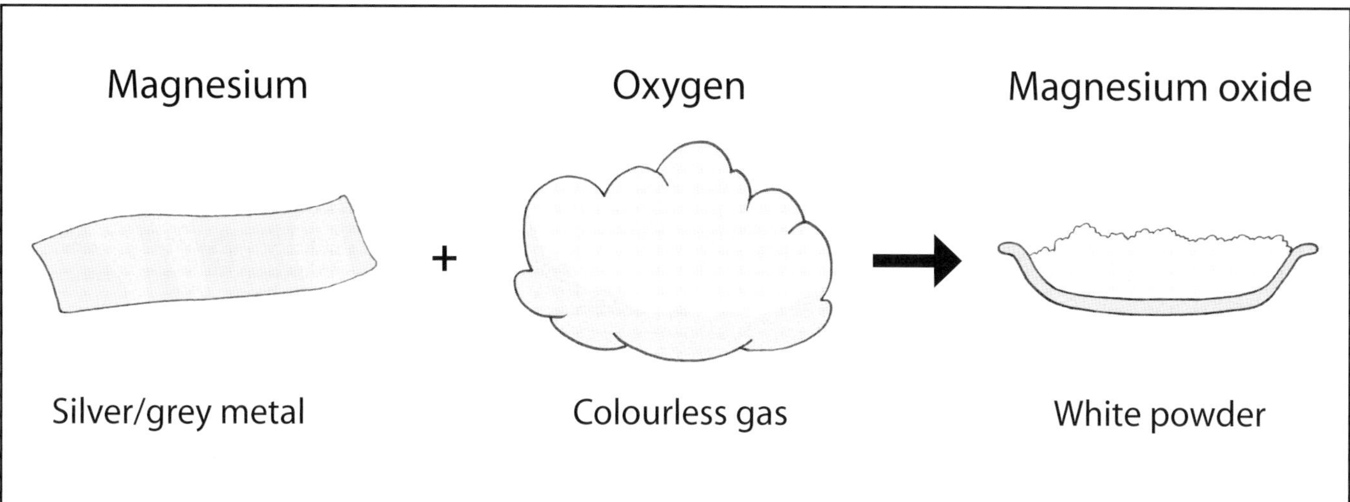

☞ Using the example above as a guide, create your own word equations for the reactions below.

- Hydrogen gas + oxygen gas → liquid water
- Iron metal + yellow sulphur powder → black iron sulphide powder
- Blue copper sulphate liqiud + iron metal → green iron sulphate liquid + copper metal
- Calcium metal + colourless water → colourless calcium hydroxide liquid + hydrogen gas (bubbles)

☞ Can you find out what salt (sodium chloride) is made from? Write down the equation in your work book, using illustrations to help you.

Element or compound?

☞ Draw arrows to show if each chemical below is a compound or an element. One has already been done for you.

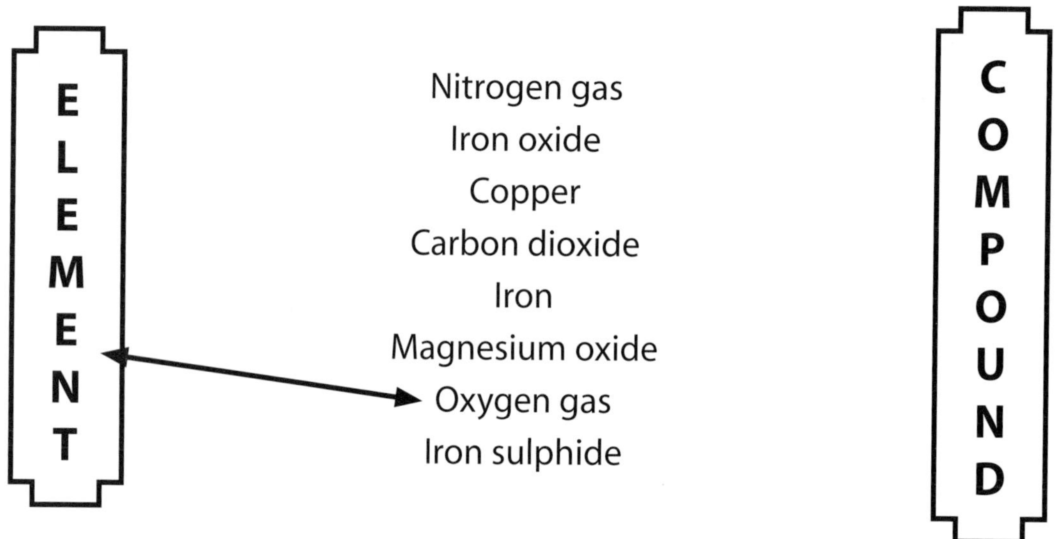

☞ Here are the names of some compounds, which you might find in a school laboratory. Can you work out which elements have been chemically joined together to make the compound? Record your answers in the spaces provided below.

Compound	Chemical formula	Number of atoms of each element
Sodium chloride	NaCl	1 Na, 1 Cl
Iron oxide	FeO	
Water	H_2O	
Carbon dioxide		1 carbon, 2 oxygen
Copper sulphate	$CuSO_4$	

 © Folens (copiable page)

The atoms in a compound

What atoms is water made from? Water is made of two different elements, hydrogen and oxygen joined together. The chemical formula for water is H_2O. This means that water has two atoms of oxygen and one atom of hydrogen.

This is how we draw the atoms in a molecule of water.

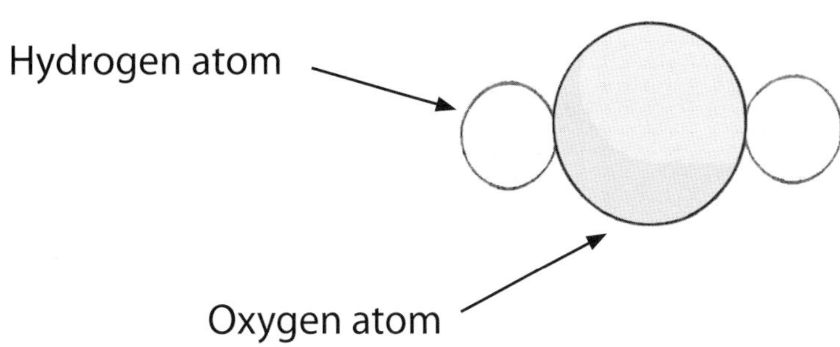

☞ Using water as an example, complete the activity, which follows.

We will use the following symbols to represent the different atoms.

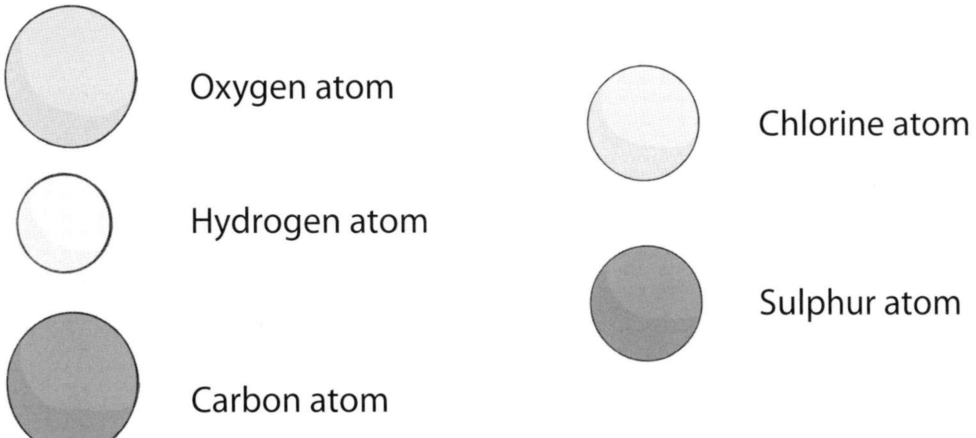

☞ On a separate piece of paper, draw a diagram to show the arrangement of the atoms in the following molecules.

A molecule of water H_2O

A molecule of oxygen gas O_2

A molecule of carbon dioxide CO_2

A molecule of hydrochloric acid HCl

A molecule of hydrogen sulphide H_2S

A molecule of chlorine gas Cl_2

☞ Write down which of these molecules are compounds.

What is mixture?

☞ Create your own word bank to complete the sentence below.

A m _ _ _ _ _ _ _ is a substance made up of atoms of two or more e _ _ _ _ _ _ _ _ or the molecules of two or more c _ _ _ _ _ _ _ _ _ that have not been chemically j _ _ _ _ _ together.

Two different elements

This is a m _ _ _ _ _ _   This is a c _ _ _ _ _ _ _

☞ Complete these sentences about different mixtures.

Air

The air around us is a m _ _ _ _ _ _ _ of different g _ _ _ _. We can s _ _ _ _ _ _ _ _ the gases from one another. The amount of each different gas in the a _ _ can vary at d _ _ _ _ _ _ _ _ _ times. Air is about four fifth's, nitrogen gas and almost one fifth o _ _ _ _ _ _ gas. There is also a s _ _ _ _ amount of argon gas and a very tiny amount of c _ _ _ _ _ _ d _ _ _ _ _ _ _ gas.

Sea water

Sea water is a mixture of s _ _ _ and w _ _ _ _.
What is the name of the process used to get pure water from sea water?

____

Fizzy drinks

Fizzy drinks are a mixture of some liquids (w _ _ _ _) and gases (carbon dioxide) and other compounds like s _ _ _ _ and flavourings.

Crude oil

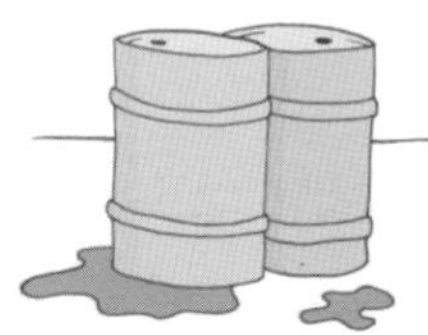

This is a mixture of lots of hydrocarbon compounds like p _ _ _ _ _ _, diesel, aviation fuel and bitumen. All the different h _ _ _ _ _ _ _ _ _ _ _ _ in crude oil can be separated out and used either as f _ _ _ or for the chemical industry.

© Folens (copiable page)

Pure substances

The word **pure** in science means a substance, which has nothing else mixed in with it. Pure water has nothing but water. Sea water is not pure water because it is mixed with salt. Every pure substance has its own special melting point and boiling point. Pure water boils at 100 °C and melts at 0° C.

☞ In pairs, complete the text below.

A mixture is not a pure substance because it contains other substances. Salt water is not a pure substance because it contains w _ _ _ _ and s _ _ _ .

☞ Will salt water boil at 100° C? Will salt water melt at 0° C? Design an experiment to find out. A diagram has been provided below as a guide. Check with your teacher before carrying out your experiment.

Pure water		**Salt water**	
Melting point =	______	Melting point =	______
Boiling point =	______	Boiling point =	______

☞ Complete these sentences, referring back to what you have learnt.

- I know that p _ _ _ substances have f _ _ _ _ melting and b _ _ _ _ _ _ _ points.
- I know that mixtures do n _ _ have fixed m _ _ _ _ _ _ _ and boiling points.
- I know that the more salt I add to the water, the h _ _ _ _ _ the boiling point of the m _ _ _ _ _ _ _ will be.

© Folens (copiable page)

Reactions of metals and metal compounds

Objectives

- Know the properties and uses of metals
- Be able to follow instructions correctly to carry out an experiment
- Identify how to put experimental procedures in the correct order

Prior knowledge

Students should know the properties of metals and what some common metals are used for. Students must be able to design and carry out their own experiment.

QCA link

Unit 9E Reactions of metals and metal compounds

NC links

Sc3 Materials and their properties 1a, 1d, 1f, 2h, 3a, 3e, 3h

Scottish attainment targets

Environmental studies – Science – Earth and space
Strand – Materials from Earth
Level E

Background

Metals are determined by their properties. Most metals exhibit all their properties, and other metals exhibit only some of these properties. For example, mercury is a liquid and is not hard. The different properties of metals make them suitable for use in different ways. Both metals and metal oxides react with acids.

Starter activity

Show students various examples of different elements. Ask students to identify which of the examples are metals, and provide a reason for their answer.

Activity sheets

'Metals'. A spider diagram shows students the different properties of metals. Ask students to copy the diagram onto a sheet of A3 paper, instructing them to complete the missing text. Students may need a word bank to help them. Students can add some pictures to their own to help them to remember the properties of metals. For example, sonorous could be illustrated by a church bell.

'Do non-metals conduct electricity?' can be used as a practical activity sheet. Ask students to look carefully at the circuit diagram provided. Students should then use the available components to construct their own circuit. Using the table provided, instruct students to connect each sample of material into the circuit using the crocodile clips. Students should record their results within the table on the activity sheet. Encourage students to use their results to answer the questions, which follow.

'Using metals'. Ask each student to study the table provided. Students should then use arrows to match each material with its correct use and property. An example has been provided. Instruct students to use the table to complete the questions, which follow. Students may need to use their own general knowledge to help them complete this activity sheet.

'Metals and acids'. **Remind students that they must wear eye protection when using acids.** Ask students to work in pairs to carry out an experiment to find out what happens when metal is added to an acid. Students should follow the instructions provided. Ask students to record their observations. Encourage students to use knowledge from a previous topic to answer the questions, which follow.

'Metal oxides and acids' is a sequencing exercise. Ask students to cut out the boxes provided and put them in the correct order. Once sentences have been placed in the correct order, ask students to construct a plan for their own experiment. Remind students they will need to have their experiments checked by a teacher before they start any practical work. Students can work with a partner to carry out their experiments. Students should then complete the equations, which follow. This activity sheet may take up two lessons.

Plenary

Use pre-made cards to match each metal property with its meaning.

 © Folens

Metals

☞ Study the spider diagram below and complete the boxes. Now, copy your diagram onto a sheet of A3 paper. Add illustrations to demonstrate the different properties of metal.

METALS

- **Ductile** Can be pulled into a w _ _ _.
- **High** m _ _ _ _ _ _ _ point.
- Good **thermal (h _ _ _)** conductor.
- **S** _ _ _ _ _ _ and can hold their shape without b _ _ _ _ _ _ _ _.
- H _ _ _ density and are **h** _ _ _ _ _.
- Good conductor of **e** _ _ _ _ _ _ _ _ _ _ _.
- All **s** _ _ _ _ _ except mercury.
- **Sonorous** r _ _ _ when h _ _ with a hammer.
- **Malleable** Can be h _ _ _ _ _ _ _ _ into a sheet and shaped.
- Shiny when **s** _ _ _ _ _ _ _ _ _.

Do non-metals conduct electricity?

☞ Study the circuit diagram below. Set up your own electrical circuit. Plan your circuit first and check with your teacher before you begin.

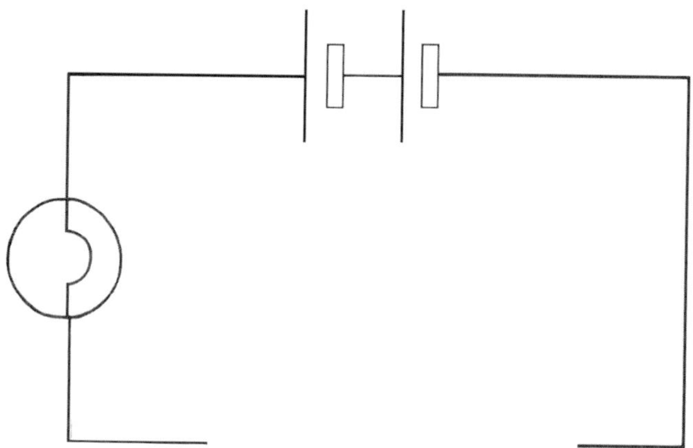

☞ Test the following metals and non-metals, which appear in the table below to see if they conduct electricity. Record your results in the spaces provided.

Material used	Metal/non-metal	Did the bulb light up?
Aluminium		
Plastic		
Rubber		
2p coin		
Iron		
Tin foil		
Cork		
Copper		

☞ What do you know?

I know that m _ _ _ _ _ are g _ _ _ conductors of electricity.

Non-metals are p _ _ _ conductors of electricity.

☞ Can you think of a way to show that metals are good conductors of heat?

____

 © Folens (copiable page)

Using metals

☞ Study the table below. Match each material with its correct use and the property, which makes it good for this use. One has already been done for you.

Material		Use		Property
Copper		Bridges and railway tracks		Lightweight and easily carried
Iron		Electrical wiring		Does not corrode easily
Aluminium		Jewellery		Strong and cheap
Gold		Ladders		Good conductor of electricity

☞ Now, answer the questions below in the spaces provided.

Copper is a metal. What properties of copper make it the best choice for electrical wiring in the home?

____

Why is the metal iron a good choice for building bridges?

____

Copper pipes can be easily bent into shape without snapping. How is this property of copper used in the home?

____

What metal is used in the manufacture of helicopters? ____________________

Why? ______________________________________

Copper saucepans are very expensive. What property of copper saucepans makes them popular with all the best chefs?

____

© Folens (copiable page)

Metals and acids

☞ Use the word bank provided to complete the text below.

Word bank

other dangerous react
metals not

Not all metals r _ _ _ _ with dilute acids in the same way. Some m _ _ _ _ _ do n _ _ react at all. But with some o _ _ _ _ metals the reaction is very violent and d _ _ _ _ _ _ _ _ !

☞ Working in pairs, find out what happens when a metal is added to an acid by completing the experiment below. **Remember to always use eye protection when handling acids.**

Apparatus

Small pieces of the metals; magnesium, zinc and copper

6 test tubes and a stand

Dilute hydrochloric acid

Dilute sulphuric acid

What to do

1. Put three of the test tubes in the stand.
2. Half fill each test tube with the dilute hydrochloric acid.
3. Add a piece of magnesium to test tube 1.
4. Repeat using zinc and copper in test tubes 2 and 3.
5. Repeat the experiment but this time use dilute sulphuric acid instead.
6. In your work books, draw what you can see in each test tube.

☞ Once you have completed your experiment, answer these questions.

1. How did you know a reaction had happened?

____

2. The gas made was hydrogen. What is the test for hydrogen gas?

____

© Folens (copiable page)

Metal oxides and acids

☞ The sentences below tell you how to make blue crystals of copper sulphate using copper oxide and dilute sulphuric acid.

The sentences have been placed in the wrong order. Cut out each sentence and stick it in your work book in the correct order. Check with your teacher before you stick each one in. Once the sentences are in the correct order, carry out the experiment.

Leave the hot liquid to cool down.	Heat the solution until about three quarters of the liquid has been evaporated.
Warm the test tube containing the mixture by putting it into a beaker of hot water.	Filter the mixture and collect the filtrate in an evaporating dish.
Leave the hot solution to cool down.	Filter again to collect your copper sulphate crystals.

☞ Complete the word equations below.

Copper o _ _ _ _ + sulphuric a _ _ _ ⟶ copper s _ _ _ _ _ _ _ + water

Zinc oxide + nitric acid ⟶ z _ _ _ n _ _ _ _ _ _ + w _ _ _ _

l _ _ _ + hydrochloric acid ⟶ l _ _ _ chloride + water.

☞ What have you learnt? Fill in the missing text below.

I know that sulphuric acid always makes a s _ _ _ _ _ _ _ .

Nitric acid always makes a n _ _ _ _ _ _ .

Hydrochloric acid always makes a c _ _ _ _ _ _ _ .

© Folens (copiable page)

Patterns of reactivity

Objectives

- Know that some metals are more reactive than others
- Understand that metals can be placed in a series according to their reactivity
- Be aware of special safety procedures when burning the alkali metals in air
- Know how to write a word equation

Prior knowledge

Students should already know the properties of metals. Students will also be able to identify that the more reactive metals are found in group 1 of the periodic table (Atoms and elements).

QCA link

Unit 9F Patterns of reactivity

NC links

Sc3 Materials and their properties 1d, 3a, 3b, 3c, 3d

Scottish attainment targets

Environmental studies – Science – Earth and space
Strand – Changing materials
Level E

Background

Some metals react more quickly when burnt in air or put into water. By comparing how different metals react with oxygen, acid, and water, we can put them in order of how quickly they react. The most reactive metal is the one that reacts most quickly and the least reactive is the one that is slowest to react.

Starter activity

Show the class a piece of sodium metal. Cut into the sodium; one student should set a stopwatch as soon as the teacher has cut into the sodium. Ask the student to stop the stopwatch when the cut sodium metal has lost its shine. Ask the students to discuss whether a shiny piece of iron would tarnish as quickly.

Activity sheets

'Metals and air'. Students will need to recall the fire triangle from 'Simple chemical reactions', whilst completing this activity. Read through the activity with students. Ask students to predict which metal will catch fire most easily, and which metal will be the slowest to catch fire. Ask students to work in pairs to complete the experiment, which follows. Students should record their results as they find them in the table provided. Refer students to their earlier predictions. Were they correct?

'Metals and water' follows on from the previous activity sheet ('Metals and air'). Teachers should carry out the demonstration provided. **Teachers should note this is a demonstration activity only and a safety sheet must be placed between the students and the experiment.** Following the demonstration, students should identify the results and record them in the table provided. Ask students to place the three metals in order of their reactivity.

'Displacement reactions'. Ask students to study the displacement reaction, which is provided. Encourage students to write their own word equations for the displacements reactions on the activity sheet. Students should then be able to place the three metals in order of their reactivity.

'Reactivity series' builds on the earlier activity sheets. Ask students to complete the word equations to see how metals react with other substances, using what they have already learnt. Give students access to text books in order to complete the activity, which follows. Instruct students to create a large version of the full reactivity series. Encourage students to add in illustrations, where they will help students to remember the order of the metals.

'Quiz time!' Divide students into four or five teams, appointing one student in each team to write down the answers. Teachers should read out the questions provided. Allow students time to discuss the answers in their teams.

Plenary

Provide students with pre-made elements cards. Ask them to arrange the cards into a reactivity series.

© Folens

Metals and air

☞ When a metal burns, it combines with the oxygen in the air to make a new substance called an oxide. An experiment has been provided below. In pairs, follow the experiment to see what happens when three different metals are burnt in air.

We will be using:
magnesium, iron, copper, tongs, a Bunsen burner, a heatproof mat, goggles.

Look through **blue** glass when burning **magnesium**.

What to do

1. Hold a small piece of magnesium ribbon in the centre of a hot Bunsen burner flame until it catches fire.
2. Write down what you see in the table below.
3. Repeat 1 and 2 above using iron and then copper.

Metal	What did the metal look like before burning?	What did you see as the metal was burning?	What did it look like after burning?
Magnesium			
Iron			
Copper			

☞ Now, complete these word equations.

Magnesium + Oxygen ⟶ Magnesium O_ _ _ _

Iron + O_ _ _ _ _ ⟶ I_ _ _O_ _ _ _

C_ _ _ _ _ + O_ _ _ _ _ ⟶ Copper O_ _ _ _

☞ Which metal burnt the brightest?

____

☞ What else was given out when the metal burnt?

____

© Folens (copiable page)

Metals and water

☞ Your teacher will show you what happens when some of the group 1 metals are put into water. Watch the demonstrations carefully and complete the text, which follows. What did you see? Record the results in the table provided.

Group 1 m _ _ _ _ _ are very r _ _ _ _ _ _ _ _ . In fact, they are so reactive that they have to be stored in oil. What would happen to these metals if they were left in air?

Metal	Symbol (Use Periodic Table)	Colour of metal	What did you see?
Lithium			
Sodium			
Potassium			

☞ Complete the findings below.

P _ _ _ _ _ _ _ _ pops and b _ _ _ _ with a l _ _ _ _ coloured flame. Moves very q _ _ _ _ _ _ across the s _ _ _ _ _ _ of the water and f _ _ _ _ _ very vigorously.

S _ _ _ _ _ moves q _ _ _ _ _ _ across the water and makes lots of b _ _ _ _ _ _ and fizzes. It ends with a yellow f _ _ _ _.

L _ _ _ _ _ _ moves across the surface of the w _ _ _ _ and fizzes making s _ _ _ bubbles.

☞ Put the three metals used in order of reactivity.

__________ __________ __________

Most reactive ← → Least reactive

© Folens (copiable page)

Displacement reactions

☞ What happens when we put iron into a solution of copper sulphate? Complete the text below.

C _ _ _ _ _ sulphate is a clear b _ _ _ solution. I _ _ _ is a s _ _ _ _ , greyish coloured metal.

☞ Now, carry out the experiment below.
Put a clean nail into a test tube half filled with copper sulphate solution and wait. After three minutes remove the nail. Write down what has happened to the nail.

____

☞ Has anything happened to the copper sulphate solution?

____

The experiment you have carried out is called a **Displacement reaction**. Iron is a more reactive metal than copper. Iron displaces (takes the place of) the copper in the sulphate solution.

☞ Now, cut out and rearrange the words below to make the word equation for the displacement of copper by iron. Stick them into your work book.

Copper sulphate	Green solution	Grey solid	Clear blue solution	+
Iron sulphate	+	iron	Brown solid	Copper	→

☞ Can you write word equations for the displacement reactions below?

- Copper sulphate and magnesium

  C __________ Sulphate + m __________ ? __________ __________ + copper

- Silver nitrate and copper

  __________ __________ + c __________ ? __________ nitrate + __________

☞ Magnesium displaces both copper and iron. Put these three metals in order of reactivity: silver, copper, magnesium.

__________ __________ __________

Most reactive ← → Least reactive

Reactivity series

☞ As we have seen, metals can react with other substances to make new compounds. Complete the word equations below. You may need to refer back to another topic to help you.

Most metals will burn in air to make a new compound called an oxide.

- Iron + oxygen ⟶ I _ _ _ o _ _ _ _
- Magnesium + O _ _ _ _ _ ⟶ M _ _ _ _ _ _ _ _ O _ _ _ _

Most metals also react with water to make either a metal oxide or a metal hydroxide.

- Iron + hydrochloric acid ⟶ iron chloride + h _ _ _ _ _ _ _ and the gas hydrogen.
- Iron + water ⟶ iron oxide + h _ _ _ _ _ _ _.
- Magnesium + w _ _ _ _ ⟶ magnesium hydroxide + h _ _ _ _ _ _ _.

Most metals react with dilute acids to make a salt and hydrogen gas. You learnt about this when you studied the topic on metals.

- Iron + hydrochloric acid ⟶ iron chloride + h _ _ _ _ _ _.
- Iron +sulphuric acid ⟶ iron s _ _ _ _ _ _ _ + h _ _ _ _ _ _ _.
- Iron + nitric acid ⟶ iron n _ _ _ _ _ _ + h _ _ _ _ _ _.

The reactivity series

By looking at the ways different metals react with air, water and acid we can make a table to put the meatals in order of their reactivity. We call this the **reactivity series**.

Make up your own reactivity series. You can use a text book to help you. Draw pictures about each metal, for example, a nail for an iron. This will help you to remember the order of the different metals in the reactivity series. Some examples are provided for you below.

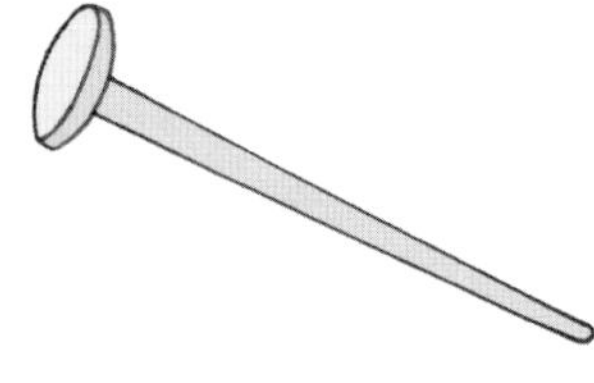

Iron nail

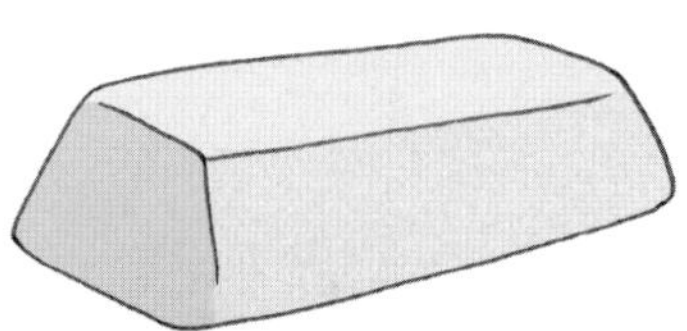

Gold bar

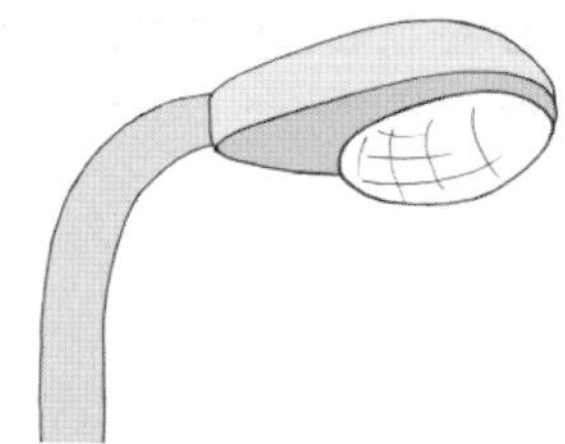

Sodium lighting

 © Folens (copiable page)

Quiz time!

☞ In teams, complete the quiz below. Cross out any answers that are incorrect.

Metals are all good conductors of	• heat and light • heat and electricity • light and electricity
Malleable means	• pull into a wire • hammer into a flat sheet • shaped like a pipe
Copper is used to make saucepans because	• it is a good conductor of heat • it conducts electricity • it can be easily shaped
When a metal is burnt in air, it always makes a new compound called	• a metalloid • a metal oxide • a metal oxygen gas
Sodium reacts with water to make the gas	• hydrogen • nitrogen • oxygen
Potassium is more reactive than lithium and sodium.	• True • False
Hydrogen gas burns with	• a purple flame • a squeaky pop • lots of sparks
Metals react with hydrochloric acid to make salts called	• nitrates • sulphates • chlorides
A more reactive metal will always displace a less reactive metal.	• True • False
All metals are magnetic.	• True • False

© Folens (copiable page)

Environmental chemistry

Objectives

- Understand that a knowledge of chemistry can help us to improve our environment
- Know the causes of acid rain and how it affects our environment
- Identify the causes of global warming and what we can do to slow it down

Prior knowledge

Students should know that the weather can affect our environment. Students also need to be aware that pollution from factories can affect the atmosphere and pollute both the land and the water. Students will already be familiar with the pH scale.

QCA link

Unit 9G Environmental chemistry

NC links

Sc2 Life processes and living things 3a, 3c
Sc3 Materials and their properties 1g, 2e, 2i, 3a, 3e, 3f, 3g

Scottish attainment targets

Environmental studies – Science – Earth and Space
Strand – Changing materials
Level E

Background

Rocks are eroded naturally to make soil. Industry and its resultant pollution affect the atmosphere. Acid rain has an adverse effect on plant life, animal populations and the quality of rivers and lakes. Pollution has also contributed to global warming and changes in sea levels, which threaten our coasts.

Starter activity

Ask students to write down what they think the word 'environment' means to them.

Activity sheets

'What is soil?' recalls ideas from unit 8G 'Rocks and weathering'. Ask students to complete the text using the word bank provided. Students should follow the instructions, to complete the experiment, which follows. The soil will take about half an hour to settle into its different layers. Ask students to draw what they see in the measuring cylinder. Tell them to describe and label each layer.

'Finding the pH of soil'. Students may need to refer to the pH scale to complete this activity sheet. Ask students to create their own word bank to complete the sentences provided. Teachers should read through the remaining activities as a class. Instructions for an experiment, which looks at the pH of soil are provided. Ask students to work in pairs to complete the prediction. Tell students to then carry out the experiment, instructing them to design their own table in which to record their results. Students should use their own results to complete the conclusion.

'Air pollution' explains to students what air pollution is. Students can carry out their own air pollution investigation using the instructions provided. Ask students to read through the instructions carefully. Students should then carry out the investigation, recording their results in the table on the activity sheet. Ask students to complete the tasks, which follow, using what they have learnt. Some students may need a word bank to help them to complete this activity.

'Greenhouse effect' provides a resource sheet for students. Ask students to study this information provided. Students should then use the information to produce a poster about global warming.

'Acid rain'. Ask students to complete the text on sheets 1 and 2 using the word banks provided. Encourage students to use the text on the activity sheet to start a class discussion, looking at how we could reduce acid rain and its effects.

Plenary

Ask each student to come up with one thing that anyone could do to prevent pollution increasing. Students should present their ideas to the rest of the class.

© Folens

What is soil?

☞ Complete the text below, using the word bank provided.

Word bank

plants dissolved sand decaying
clay water animals mixture
earthworms rock

Soil is a m _ _ _ _ _ _ of different things.
It contains:

- bits of r _ _ _ , s _ _ _ and c _ _ _
- humus which is made from the dead and d _ _ _ _ _ _ _ remains of p _ _ _ _ _ and a _ _ _ _ _ _
- air and w _ _ _ _ are in the spaces between soil particles
- some d _ _ _ _ _ _ _ _ chemicals
- living animals like e _ _ _ _ _ _ _ _ _ and parts of plants like roots (carrots) and potatoes.

☞ Complete the following experiment using the instructions below.

Put about 30cm^3 of soil into a measuring cylinder. Cover the soil with 50cm^3 water. Shake carefully to mix the soil and water together. Leave to settle. Now, draw what you see on the measuring cylinder opposite and label each layer.

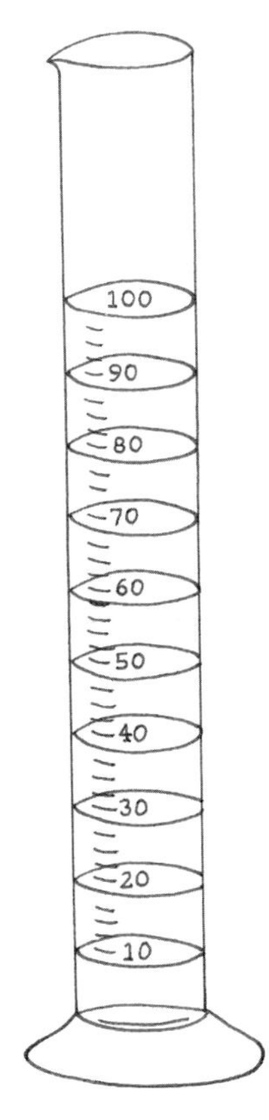

© Folens (copiable page)

Finding the pH of soil

☞ Complete the text below. You can look back to other activity sheets to help you.

- I know that the pH scale tells me how a _ _ _ or how a _ _ _ _ _ _ a substance is.
- I know that I can use U _ _ _ _ _ _ _ _ _ Indicator to find out the pH of a substance.
- I know that with this indicator, a neutral substance is g _ _ _ _ _ .
- An acid substance is r _ _ .
- An alkali substance is p _ _ _ _ _ _ .

☞ Study the **aim** below. Work in pairs to complete the **prediction**, which follows.

Aim (What I am going to find out):
To find the pH of different soil samples.

Prediction:
I think that different types of soil will have d _ _ _ _ _ _ _ _ _ pH values.

☞ Now, plan your experiment. Use the information below to help you. Before you start, draw a results chart in your work books to show; type of soil, colour of universal indicator and pH of soil.

Plan *(What I am going to do)*:
1. Put half a spatula of the soil sample A (sandy soil) in a boiling tube and add 5cm^3 of tap water.
2. Stir the mixture.
3. Filter and collect the filtrate (liquid that comes through the filter paper) in a beaker.
4. Test the filtrate using Universal Indicator liquid.
5. Record the colour on your results chart.
6. Repeat with soil sample B (clay soil).

☞ I have found out that sandy soils are ***more/less*** acid than clay soils.

© Folens (copiable page)

Air pollution

We need to breathe in air to stay alive. How clean is the air we breathe? Breathing in dirty or polluted air can lead to illnesses like asthma. The main causes of air pollution are factories and transport. Wherever fuel is burnt, waste gases and dirt will be put back into the atmosphere.

☞ **Air pollution investigation.**

Take three leaves from a tree in three very different environments, rub the upper surface of the leaf with a clean cotton wool ball. How much dirt is on your cotton wool ball? Record your findings in the table provided below.

Source of leaf	Stick used cotton wool ball here	How much pollution?
Close to a busy road		
A quiet country road		
A field or garden		

☞ Where does this dirt come from?

___

☞ What can we do? Complete the text below, using what you have learnt.

- Use buses and trains because they carry m _ _ _ people in each j _ _ _ _ _ _ and this cuts d _ _ _ pollution.
- Walk or c _ _ _ _, this does not cause any pollution and its h _ _ _ _ _ _.
- Encourage drivers to go more slowly, it causes l _ _ _ pollution and is s _ _ _ _.

© Folens (copiable page)

Greenhouse effect

There are more and more greenhouse gases being made and put into the atmosphere. More heat is being trapped and this is making the Earth warmer. This is called **Global warming**.

☞ Read carefully through the information provided. Use the information to create a poster about global warming.

When you create your poster, think about:

- how greenhouse gases are made
- how the world will be affected
- what we can do to reduce climate change.

Greenhouse effect

The Sun's rays warm the Earth. Gases in the Earth's atmosphere trap some of the heat and stop it escaping into space.

Greenhouse gases

Water vapour is a natural atmospheric gas.
Carbon dioxide is made when plants and animals respire. Erupting volcanoes also make carbon dioxide.
Methane is a gas which is made by cattle as they digest their food. Methane also comes from rice as it grows in paddy fields.

Activities that increase greenhouse gases

Burning fossil fuels releases carbon dioxide gas.

CFCs are gases used in aerosols such as hairsprays, fridges and making foam. They are dangerous because they can trap large amounts of heat.

Effects of global warming

Weather all over the world will change. Summers will get hotter and winters will get colder.

Some places will get drier and have water shortages. Other places will flood more easily.

Changes in weather conditions affect the crops we can grow. It has taken millions of years for plants and animals to adapt to their environment. Many will not be able to adapt to the new conditions and will not survive

Ozone

This occurs naturally in the atmosphere.

© Folens (copiable page)

Acid rain sheet 1

☞ What causes acid rain? Complete the text below, using the word bank provided.

Word bank

gas electricity mix chemicals
car air oil fumes sulphur
coal power nitrogen volcanoes

When fossil fuels like o _ _ , c _ _ _ and natural g _ _ burn, they release c _ _ _ _ _ _ _ _ into the air. Two of these chemicals are s _ _ _ _ _ _ and n _ _ _ _ _ _ _ . These chemicals m _ _ with the water in the a _ _ and form oxides. Sulphur comes from p _ _ _ _ stations, which make e _ _ _ _ _ _ _ _ _ _ , and from erupting v _ _ _ _ _ _ _ _ . Nitrogen oxides come mainly from c _ _ and lorry exhaust f _ _ _ _ .

☞ This is how acid rain is made. Complete the sentences.

1. Sulphur gas + oxygen in the air ⟶ sulphur dioxide gas.
2. Sulphur dioxide gas + water vapour in the air ⟶ sulphuric acid.
3. N _ _ _ _ _ _ _ gas + o _ _ _ _ _ in the a _ _ ⟶ nitrogen oxides.
4. N _ _ _ _ _ _ _ o _ _ _ _ _ + water v _ _ _ _ _ in air ⟶ nitric a _ _ _ .

Acid rain sheet 2

☞ Use the word bank provided to complete the text below.

Word bank

winter speeds factories acid fish fizzes sea Sweden grow old erosion limestone soil leaves nutrients reduce disappear

Pollution from factories in England is blown across the s _ _ to Sweden where it falls onto the forests and lakes as a _ _ _ rain. Most of the acid rain that falls on Canada starts as pollution from f _ _ _ _ _ _ _ _ across the border in the USA.

☞ What does acid rain do? Use the same word bank to complete the text below.

Acid rain takes away minerals from the s _ _ _ so that trees do not have the n _ _ _ _ _ _ _ _ they need to g _ _ _ properly. Trees loose their l _ _ _ _ _ and become too weak to survive harsh w _ _ _ _ _ weather.

When it falls on lakes, acid rain kills plants and f _ _ _. Many of the lakes in S _ _ _ _ _ no longer have any fish left in them. Buildings made from softer materials like sandstone and l _ _ _ _ _ _ _ _ _ are gradually being eroded (worn away) by the effects of wind and frost.

Acid rain s _ _ _ _ _ up the process of erosion and on some o _ _ buildings and churches many of the statues have lost faces and limbs due to e _ _ _ _ _ _ by acid rain.

☞ Can you remember what happens when acid is put onto some chalk?

It f _ _ _ _ _ and if you keep adding more and more acid, the chalk will eventually d _ _ _ _ _ _ _ _ _. This is what will happen if we do not r _ _ _ _ _ pollution of the atmosphere.

☞ In groups, discuss how we can reduce acid rain and its effects. Present your findings to the class.

 © Folens (copiable page)

Using chemistry

Objectives

- Know what happens when a fuel burns
- Be able to distinguish between the usefulness of different fuels
- Know that other chemical reactions also produce heat
- Understand the theory of Conservation of mass and be able to give examples

Prior knowledge

Students will have a knowledge of exothermic reactions. They should know how fossil fuels were formed. Students will also be aware of the standard tests for carbon dioxide gas, hydrogen gas, and water vapour.

QCA link

Unit 9H Using chemistry

NC links

Sc2 Life processes and living things 5a
Sc3 Materials and their properties 1e, 1f, 2a, 2g, 2h, 2i, 3b
Sc4 Physical processes 5a

Scottish attainment targets

Environmental studies – Science – Earth and Space
Strand – Changing materials
Level D, F

Background

A fuel is a substance, which burns to release energy. Burning is a chemical reaction and the products include the gases carbon dioxide and water vapour. Different fuels are used in different situations. Other chemical reactions also give out heat (exothermic). During a chemical reaction, the combined mass of the reactants is the same as the mass of all the products. This is called the Conservation of mass.

Starter activity

Ask students to write down what test they would do if the gas was carbon dioxide.

Activity sheets

'What happens when a fuel burns?' introduces a practical demonstration, in which natural gas is burnt and the products are tested. Ask students to observe the experiment very carefully. Tell students to explain what is happening in each part of the apparatus. Instruct students to complete the text provided. Students should answer the questions provided on completion of the experiment.

'Combustion – another name for burning' looks at the properties of different common fuels. Ask students to complete the introductory text. Students should then work in groups of four or five to complete the activity, which follows. A spider diagram has been provided. Ask students to discuss each question, which appears within the diagram in their groups. Instruct each group to create further spider diagrams relating to the different fuels provided, using the example given. Students should then answer the questions about each fuel as before. Encourage each group to produce a diagram for at least three of the fuels provided. Students may wish to add pictures to make their answers clearer.

'Energy from chemical reactions'. Students may wish to refer back to the 'Reactivity series of metals' when completing this activity. An aim for an experiment has been provided. Ask students to form pairs. Each pair should read through the aim carefully. Instruct each pair to plan their own experiment using the aim as their guide. Ask students to draw the apparatus they will use before they begin the experiment in their work books. Tell students to follow their own plan and record their results in the chart provided. Students should use their results to answer the questions, which follow.

'Conservation of mass'. Ask students to read through the activity sheet carefully and to complete the missing text. Some students may need a word bank to help them. Instruct students to work with a partner to carry out the experiments outlined. Emphasise to students the importance of weighing materials accurately. Using the diagram provided, ask students to complete an equation for dissolving salt.

Plenary

Ask students to identify a fuel and explain where it comes from. For example, oil was made millions of years ago from dead sea creatures.

© Folens

What happens when a fuel burns?

☞ Your teacher will demonstrate what happens when natural gas is burnt. When you have seen the demonstration, complete the text below.

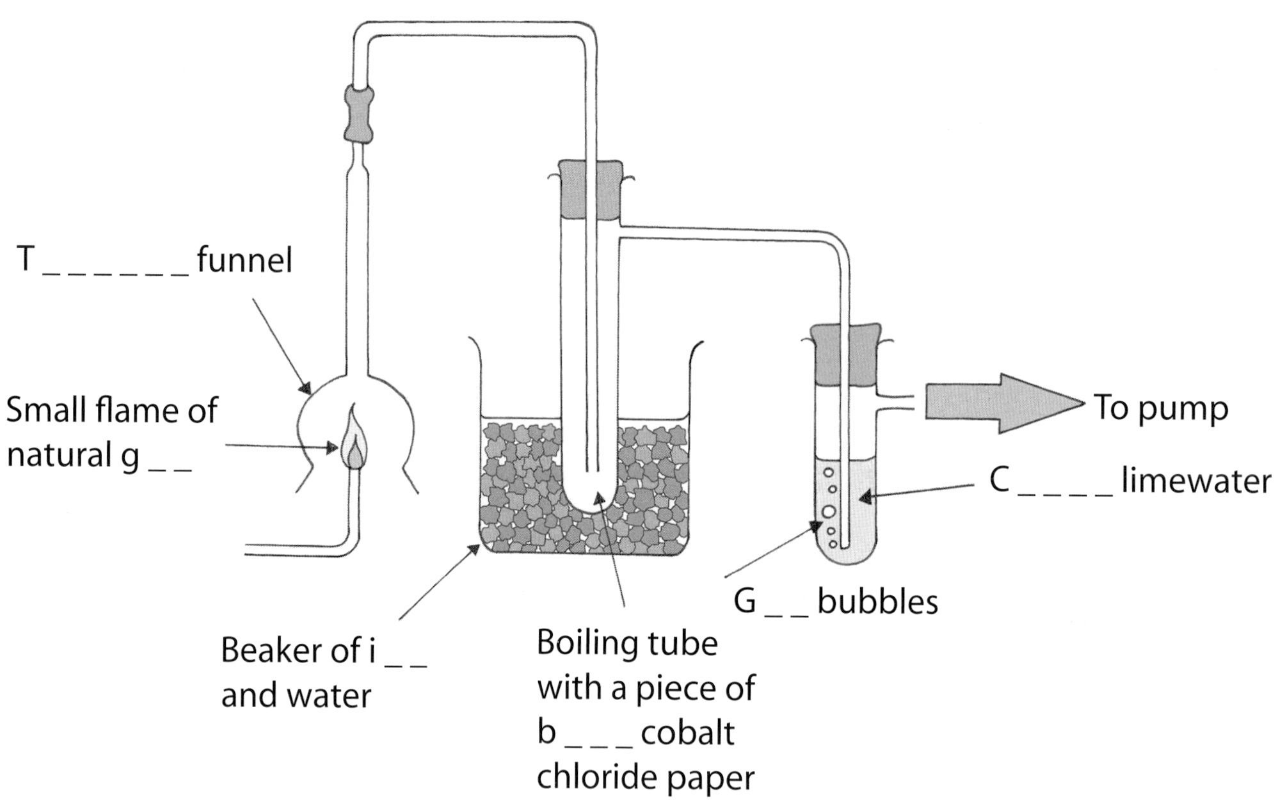

☞ What did you see happen? Using what you have learnt, fill in the missing text below.

After a few minutes we could see some b _ _ _ _ _ _ rising up through the limewater. Some droplets of a colourless l _ _ _ _ _ also collected in the boiling tube. As we watched the piece of c _ _ _ _ _ chloride paper gradually changed c _ _ _ _ _, it changed from b _ _ _ to p _ _ _.
Then we saw that the l _ _ _ _ _ _ _ _ had started to turn a m _ _ _ _ white colour.

☞ What made the cobalt chloride paper turn pink?

____

☞ What gas turns clear limewater milky white?

____

 © Folens (copiable page)

Combustion – another name for burning

☞ Complete the text below.

Natural gas or methane is a fuel. It is a compound made from carbon and hydrogen atoms only. This type of chemical compound is called a hydrocarbon. When a hydrocarbon burns it makes h _ _ _, c _ _ _ _ _ d _ _ _ _ _ _ gas and w _ _ _ _ vapour. All fuels need a good supply of o _ _ _ _ _ to burn.

Methane + o _ _ _ _ _ → w _ _ _ _ vapour + c _ _ _ _ _ dioxide + h _ _ _.

☞ Now, in groups, study the spider diagram below. Discuss each of the questions in your groups.

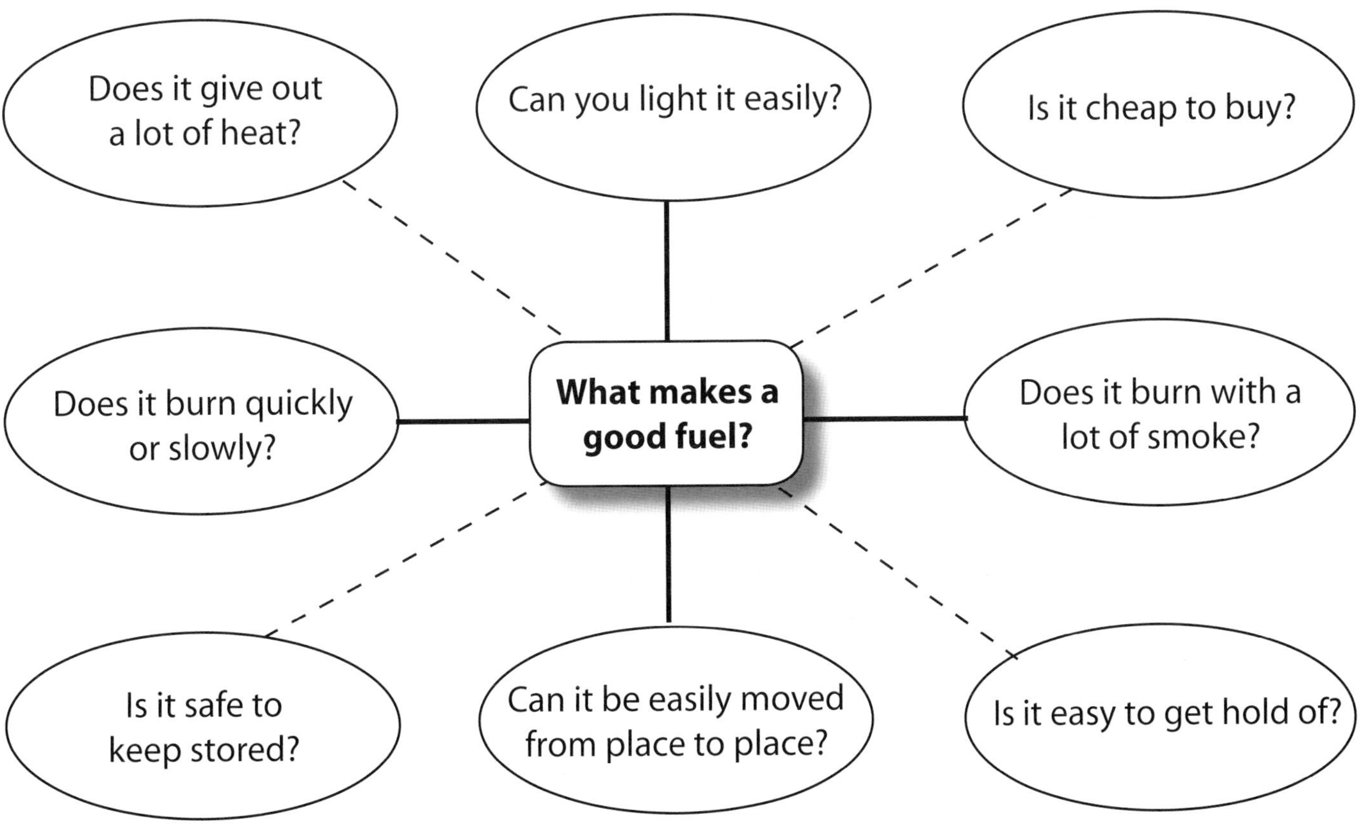

☞ Using the spider diagram above as an example, create similar diagrams for at least three of the following fuels. Answer the same questions as before for each of your chosen fuels.

Petrol *Wood* *Natural gas* *Oil* *Coal* *Peat*

© Folens (copiable page)

Energy from chemical reactions

Remember when we learnt about displacement reactions? We found out that a more reactive metal could displace (take the place of) a less reactive metal. When the reaction happens, some heat is also released. Complete these tasks in your work books.

☞ **Aim** (what I am going to find out)

Work with a partner to design an experiment to show that heat is released by these two chemical reactions.

1. Magnesium + copper sulphate ⟶ magnesium sulphate + copper.
2. Zinc + copper sulphate ⟶ zinc sulphate + copper.

☞ **Plan** (What I will do)

Now, make a list of each stage of your experiment in the order you will do it.

☞ **Equipment**

List all the apparatus you will need for your experiment. Draw a diagram of the apparatus you are using. Label all the parts.

☞ **Results**

Now, carry out your experiment. Your teacher will supervise you. Record your results in the table below.

Experiment	Temperature of copper sulphate solution at the beginning of experiment	Temperature of copper sulphate solution at the end of experiment	Rise in temperature
Magnesium + copper sulphate	______°C	______°C	______°C
Zinc + copper sulphate	______°C	______°C	______°C

☞ I found out that heat ***was/was not*** released during these two experiments.

☞ Which experiment released the most heat energy?

____

 © Folens (copiable page)

Conservation of mass

☞ Mass is neither created nor destroyed in a **chemical reaction**.

Work with a partner to complete the experiment below. Use what you find out to complete the text, which follows.

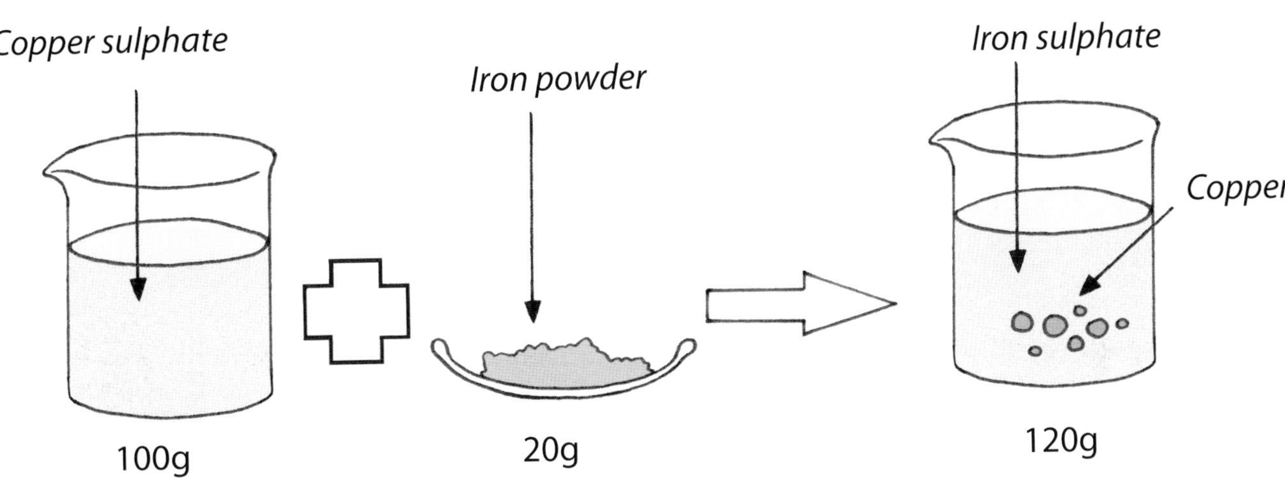

When iron powder is added to copper sulphate solution, a chemical reaction happens. The iron (more reactive metal) displaces the copper to make iron sulphate solution and copper metal.

Iron + copper sulphate ⟶ iron sulphate + copper
20g + 100g ⟶ 120g

During a c _ _ _ _ _ _ _ _ reaction atoms are n _ _ _ _ _ made nor destroyed. Atoms can only be rearranged. Nothing is u _ _ _ up and n _ _ _ _ _ _ is created.

☞ The rules are the same for a **physical reaction**.

Complete the experiment below. Write down your own equation for dissolving salt.

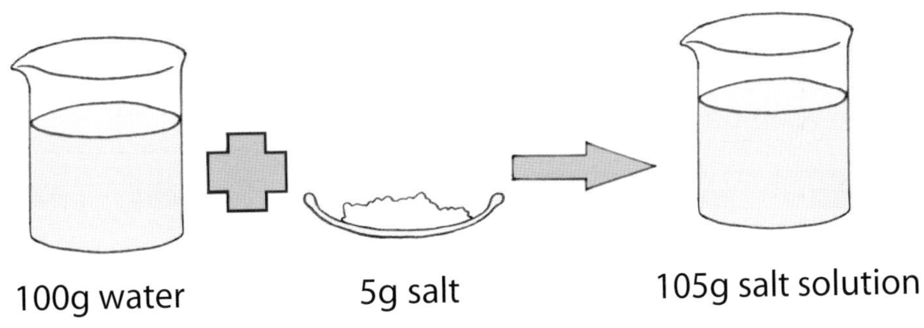

What can I do?

What I can do	Example from my work
I can measure volume accurately.	
I can select the correct apparatus for an experiment.	
I can write a plan for an experiment.	
I can make a prediction.	
I can design a results chart.	
I can use the correct units.	
I can separate substances by filtering.	
I can use chromatography.	
I can test for hydrogen gas, carbon dioxide gas and oxygen gas.	
I know the symbols for the first 20 elements in the Periodic Table.	

THE LIBRARY
TOWER HAMLETS COLLEGE
POPLAR HIGH STREET
LONDON E14 0AF

© Folens (copiable page)